TSCA's Impact on Society and Chemical Industry

TSCA's Impact on Society and Chemical Industry

George W. Ingle, EDITOR

Chemical Manufacturers Association

Based on a symposium jointly
sponsored by the ACS Divisions of
Industrial and Engineering Chemistry,
Chemical Information, Organic
Coatings and Plastics Chemistry, Small
Chemical Businesses, and the Board
Committee on Corporation Associates
at the 182nd National Meeting
of the American Chemical
Society at Las Vegas, Nevada,
March 31–April 1, 1982

ACS SYMPOSIUM SERIES **213**

AMERICAN CHEMICAL SOCIETY

WASHINGTON, D.C.   1983

Library of Congress Cataloging in Publication Data

TSCA's impact on society and chemical industry.
  (ACS symposium series; 213)

  "Based on a symposium sponsored by the ACS
Division of Industrial and Engineering Chemistry at
the 182nd national meeting of the American Chemical Society at Las Vegas, Nevada, March 31–April
1, 1982."

  1. Chemical industries—Law and legislation—
Economic aspects—United States—Congresses. 2.
Hazardous substances—Law and legislation—Eco-
nomic aspects—United States—Congresses.
  I. Ingle, George W. II. American Chemical Society.
Division of Industrial and Engineering Chemistry.
III. Title: T.S.C.A. IV. Title: Toxic Substances Con-
trol Act. V. Series.

HD9651.5.T8   1983   344.73'0424   83–2733
ISBN 0–8412–0766–6 347.304424 ACSMC8 213
1–249   1983

ACS Symposium Series

M. Joan Comstock, *Series Editor*

FOREWORD

The ACS SYMPOSIUM SERIES was founded in 1974 to provide
a medium for publishing symposia quickly in book form. The
format of the Series parallels that of the continuing ADVANCES
IN CHEMISTRY SERIES except that in order to save time the
papers are not typeset but are reproduced as they are submitted by the authors in camera-ready form. Papers are reviewed under the supervision of the Editors with the assistance
of the Series Advisory Board and are selected to maintain the
integrity of the symposia; however, verbatim reproductions of
previously published papers are not accepted. Both reviews
and reports of research are acceptable since symposia may
embrace both types of presentation.

CONTENTS

PREFACE

A COMPREHENSIVE HISTORY of the Toxic Substances Control Act (TSCA) and its impacts has not been compiled. Because these impacts will require substantial time to identify and evaluate, such a history may not be written for another decade or more. Meanwhile, detection and analysis of the effects of TSCA from a variety of viewpoints will help to delineate the beneficial and the detrimental consequences of this law. This preliminary review was the purpose of the American Chemical Society symposia on the impacts of TSCA on society and the chemical industry, September 11 and 12, 1979, at Washington, D.C., and March 31 and April 1, 1982, at Las Vegas, Nevada. Although the first of these symposia was more predictive than factual, both have indicated that the effects of such a far-reaching and complex law that affects the fourth largest U.S. industry will take much more time to understand and evaluate.

From research and development through production and disposal of chemical substances, TSCA touches on most aspects of the chemical industry. For this reason, all members of the American Chemical Society, whether in industry, education, or government, should be aware of the interaction between this law and their vocations and careers.

For their contributions of time, talent, and experience in describing these effects, the authors of these papers should be greatly applauded. To the reviewers of these papers, whose comments were taken by the authors to improve content and interpretation, my thanks are given.

The other ACS divisions that served as joint sponsors with the Division of Industrial and Engineering Chemistry and the chairmen provided for three of the four sessions of this symposium should be identified for their helpful assistance: Howard M. Peters from the Division of Chemical Information (Chemistry and the Law Subdivision); Kenneth W. Greenlee from the Division of Small Chemical Businesses; and Lawrence Keller from the Division of Organic Coatings and Plastics Chemistry. Lawrence Keller served also as co-chairman and helped develop the general scope of the symposium. The Board Committee on Corporation Associates also was a joint sponsor.

For background information on this subject, one may consult the following: (1) Annual Reports of the Council on Environmental Quality;

(2) related publications of the National Academy of Sciences; and (3) "The Business Guide to TOSCA, Effects and Actions," G. S. Dominguez, John Wiley and Sons (1979).

Finally, I must acknowledge the frequently requested and helpful guidance and advice given by W. Novis Smith, Program Chairman, and Robert A. Ference, Program Secretary, for the Division of Industrial and Engineering Chemistry.

GEORGE W. INGLE
Chemical Manufacturers Association
Washington, DC 20037

December 1982

Background, Goals, and Resultant Issues

GEORGE W. INGLE

Chemical Manufacturers Association, Washington, DC 20037

At least two parts of the history of TSCA have been prepared. The legislative history was prepared by the Library of Congress, Congressional Research Service, shortly after then President Ford signed the bill into law, as Public Law 94-469. A significant part of this summary is Appendix I, of the April 1971 report, "Toxic Substances" prepared by the Council on Environmental Quality (CEQ). The nearly six years of legislative activity began when then President Nixon included the essence of this report in his State of the Union Message to Congress in February 1971.

While this CEQ report is the legislative origin of TSCA, the conceptual origin, like that of each of the several pieces of environmental legislation beginning in the mid-sixties, may well be Rachel Carson's "The Silent Spring," published in 1962. Several other related and heavily publicized events created intensified interest; these involved vinyl chloride monomer (VCM), polychlorinated biphenyl (PCB's), mercury and other substances associated with biological damage.

A second part of the history of TSCA was prepared by the Chemical Manufacturers Association (CMA) -- "The First Four Years of the Toxic Substances Control Act -- A Review of the Environmental Protection Agency's Progress in Implementing TSCA." This review completed and summarized the "significant developments in the interpretation and implementation of TSCA since its enactment and CMA's assessment of them."

There seems to be no corresponding analysis by the initial proponents of TSCA, the group of environmentalist organizations including the Natural Resources Defense Council, the Environmental Defense Fund, The Conservation Foundation, and others. In time, one expects that such a perspective will be contributed.

Each of these different views should be of concern to the broad spectrum of members of the American Chemical Society. It is their disciplines and their industry which are or will be affected in some way by the concepts, procedures and controls in TSCA.

0097-6156/83/0213-0001$06.00/0

While the entire CEQ Report of April 1971 should be read, its conclusions should be stressed:

1.   Toxic substances are entering the environment and these substances can have severe effects;

2.   Existing legal authorities are inadequate and new legal authorizations are required. Those authorities included in the President's February '71 report were:

-- EPA's authority to restrict or prohibit use or distribution of a chemical substance, to protect health or the environment; not only adverse effects but desired benefits must be considered;

-- If the hazard were imminent, EPA could ask the courts to restrain use or distribution of the substance immediately;

-- EPA would be authorized to issue standards for tests to be performed and for results to be achieved for new substances, which could be marketed only after meeting these standards;

-- EPA could request from manufacturers information on potentially toxic substances -- names, composition, production level, uses, and results of tests to evaluate their effects;

-- The Council on Environmental Quality would be charged with coordinating efforts to establish a uniform system for classifying and handling information on chemical substances.

It was further concluded that the Toxic Substances Control Act is a new way of looking at environmental problems, a systematic and comprehensive approach, not limited to pollutants classified by their occurrence, as in air or water. TSCA contemplates the flow of potentially toxic substances from their origin, through use, to disposal.

In the five and one-half years of ensuing Congressional activity, many additional aspects were considered and some were included in the Act as finally enacted. Possibly the most controversial had to do with the treatment of new substances. Were these to be treated by registration, as is the case in the Federal Insecticide Fungicide and Rodenticide Act (FIFRA), or were they to be subject to a less onerous notification procedure? This would begin the attempt to assess their risks more in balance with the growth in commercial volume of the substance, and hence with its capacity to pay the costs for the frequently costly testing required. Without such a balance, the

industry asserted that excessive costs for testing new substances without established markets would frustrate their
research and development. The notification view prevailed
finally in the form of a flexible and sequential review of new
substances, and of existing substances, including those new
substances found to present no unreasonable risk and thus, in
time, added to the inventory of existing substances.

This decision is the source of several problems discussed-
in this symposium. Does this multistage system of assessing the
risks of new substances simultaneously protect and nurture
chemical innovation, as Section 2 of TSCA includes as part of
Congressional Policy and Intent? The papers by D. Bannerman and
C.W. Umland deal explicitly with this issue and others
indirectly. Has the European Economic Community's Sixth Amendment to its June 27, 1967, Directive (relating to the
classification, packaging and labeling of dangerous substances)
created an international impasse by establishing, for notifying
new substances, a system nearer to registration? B. Biles's
chapter suggests this has happened, and that it will take these
groups of trading partners years to resolve. This problem is
aggravated further by OECD's (Organization for Economic Cooperation and Development) proposal of its "Minimum Premarketing
Data" requirement for assessing the risks of new chemicals.
Almost identical to the EEC's "base set" of data for the same
purpose, these two rigid systems are difficult to harmonize with
TSCA on a sound risk-assessment basis, but the effort will
continue.

Another balance of factors within TSCA has to do with the
risks and benefits of the far larger number of existing chemicals. E.H. Blair's contribution examines the problem of setting
priorities for testing exsiting chemicals to assess their risks
in a cost-effective procedure.

Of all that which is known about the risks of these
substances to health and to the environment, how much is significant? What are the further needs for information? How much
of this may exist in unpublished work elsewhere in the new
world? How can unnecessary duplication be eliminated? How far
should the OECD Chemicals Programme go in internationalizing
review and evaluation of existing chemicals for their risks? Is
it true, as E. H. Hurst asserts in his chapter, that the costs
of notifying new chemicals are so great, in relation to their
commercial value, that USA research and development increasingly
examines existing substances, or closely related new substances,
with minimal risks? These questions should be pondered by
chemists in research and development, because they bear on

future of the chemical industry, here and abroad, and thus on chemists' careers.

These effects and those of other regulations proposed or implemented by EPA have stimulated a flow of initiatives by the chemical industry and its major trade associations to propose changes in these or new concepts for other, regulations, as described by S. Davis, Esq., in her analysis. Many of these changes and concepts are of particular interest to the smaller chemical manufacturer. Their limited financial and manpower resources are far less able to cope with the requirements of TSCA, with the result that this significant source of chemical innovation is at a serious disadvantage.

In its full reach, TSCA requires the receipt, production and management of a vast amount of information. As C. Elmer and D. Harlow indicate, major new incentives for larger and better information management systems within the corporate structures of most chemical manufacturers have resulted. The role of EPA itself, in amassing such information, maintaining its quality, and making it available within the limits of confidentiality controls, needs critical examination. The significance of those elements of this information that are trade secrets is discussed by J. O'Reilly. He stresses significant differences between the untried EEC system under the Sixth Amendment to the 67/548/EEC Directive, and the functioning TSCA System -- differences which need resolution. The utility of TSCA's information banks to people in chemical market research, for example, was not described in this Symposium, but it would seem to be only a matter of time before this major compilation will attract workers in this and other parts of the chemical industry, as well as workers in the government itself and people representing public interests.

The impacts of TSCA, such as those on two specific exemplary industries, surface coating polymers and metal-cutting fluids, by S.Oslosky and H.Fribush, respectively, are implied but actually not explicit within TSCA. Consider the required assessment of risks, the need for test-data describing effects on health and the environment, aside from plant inspections, subpoenas, prohibited acts, penalties for prohibited acts, enforcement and seizure, judicial review, citizens' civil actions and petitions, and employee protection provisions in the Act. Thus, it's inevitable that the alert manufacturer will adjust his product research, development and selection processes to identify and use substances with reduced risk to health and the environment wherever possible. As structure-(biological)-activity relationships become more reliable, the alert

synthesizer of new substances should rely increasingly on this
discipline to help reduce costs in sharpening his selection of
preferred structures. These factors direct the chemical industry
toward substances, uses and controls which should be generally
preventive of insult to health and environment.

Aside from the actions already initiated by EPA under Section 6
to restrict exposures to polychlorinated biphenyls and to
chlorofluorocarbons in certain uses, no other actions have been
taken against specific chemical substances, nor has an imminent
hazard been identified for appropriate action. Less than a dozen
proposed orders have been issued under Section 5(e) requesting
further information to assess the risks of as many new
substances. Perhaps 80 informal requests for further information
on such substances have been made and satisfied voluntarily.
Testing programs for a substantial number of existing substances
have been started and more are planned. In addition, of course,
the monumental task of creating an inventory of some 55,000
existing chemicals was completed.

With very few short-term exceptions, these actions may
lead hopefully to long-term improvement, largely by preventive
measures, and by broad education of manufacturers, processors
and users, to reduce health and environmental insults. How can
such meaningful progress, if any, be measured, aside from
numbers of regulations issued, or chemical compounds tested? How
can it be determined if this noble experiment is successful, let
alone cost-effective? M.J. Lipsett's paper suggests that impacts
of TSCA on public and occupational health may take a long time
to detect, if ever, simply because TSCA is only one in a
spectrum of related laws.

How, then, can decisions be made under TSCA if the effects
are so difficult to discern? Some insight into useful and
quantitative methodology is given by D.W. North.

Equally or more important for the long term, is the supply
of talent to use this or other methodology to make such
decisions, whether in the regulatory agencies or in the
industry. To the extent that relevant sectors of industry take
the full range of initiatives to reduce adverse biological
effects, aside from complying with existing and forthcoming
regulatory requirements -- the regulatory agencies' roles may be
minimized. To this essential goal, R.L. Perrine's comments on
educating the Ph.D. environmental chemist for careers in
government, industry, and in education are directed.

Having asked all these questions, and made analyses and
drawn conclusions from observing implementation of TSCA to date,
how can one sum its overall costs and its benefits? On balance,

do the benefits equal the costs? The costs are consistently more evident than the benefits, but is this simply because the latter are so difficult to quantify? Much information has been generated about new and existing chemical substances, but EPA's control of unreasonable risks related to these has been very much less in evidence. Does this reflect the use of other laws, EPA's inaction, industry's self-control or industry's opposition to any control by EPA? The Conservation Foundation's J.C. Davies, who had a significant role in CEQ's initial creation of TSCA, commented, "there have been almost no significant benefits of TSCA resulting from controls on existing chemicals. The costs of administration of TSCA in '81 approximated $85 million, and industry and public costs for compliance in '80 were estimated by CEQ to be roughly 3.5x this level."

Are the indirect costs more important than these direct costs? Has innovation in the chemical industry been supressed by TSCA? How reliable are the assumptions that such suppression has occurred? Can TSCA's effects be separated from those of inflation, tax rate, industrial R/D budgets, and long-term maturation of the industry?

Davies suggests that the uncertainties in costs and in benefits are so great as to frustrate their comparison. Aside from this is the philosophy that the public, through TSCA, has a role in decision-making with regard to chemicals. In this context, the public might well ask if experience to date in implementing TSCA indicates areas of inefficiency. Are all parts of TSCA workable? What is the most effective balance between regulatory action and voluntary compliance? Does TSCA provide a "due-process" for industry and the public in determining if a substance poses unreasonable risks? Does TSCA gather more information than is needed for appropriate control of chemicals? Does TSCA provide a reasonable framework for deciding que_ ions of policy, if not of science, as to how much risk society will tolerate in using chemical substances?

Few of these questions are easy to answer. The effort has begun in these papers. All members of the American Chemical Society should be concerned that progress is made toward their answers.

RECEIVED August 10, 1982

Impact on Market Introduction of New Chemicals

DOUGLAS G. BANNERMAN [1]

U.S. Environmental Protection Agency, Washington, DC 20460

The Toxic Substances Control Act (TSCA) refers to "new chemicals" as those not on the TSCA Inventory of Chemical Substances which lists about 55,000 existing commercial chemicals. All new chemicals must enter EPA's premanufacture notification program (PMN) for review before manufacture. This program is the most complete record of development of new chemicals by U.S. industry over the past 2 1/2 years. To date over 1,000 notices have been submitted, many including confidential business information (CBI). Despite the CBI, it is possible to summarize EPA's experience with new chemical substances and to evaluate the PMN program and its impact on product innovation. That is essentially the aim of this paper.

Assessments of these new chemicals are made by teams of multidiscipled scientists, and are based on limited firm data, comparisons to similar chemicals and estimations of exposure to humans and the environment. Generally, these PMNs contain some information on acute health effects but relatively scant information on chronic health and environmental effects. Although there is no way of knowing how many harmful chemicals were kept off the market since this PMN requirement went into effect, this risk assessment process was never applied so uniformly and thoroughly in pre-TSCA days and in this respect TSCA is meeting one of its major goals.

Information presented and discussed includes the number of PMNs submitted, an analysis of the classes and types of new chemicals, the most active product development areas and the actual number of new chemicals which have been commercialized after

[1] Current address: National Electrical Manufacturers Association, Washington, D.C. 20036

0097-6156/83/0213-0007$06.00/0
© 1983 American Chemical Society

completing the review process. Intermediates in the
manufacture of other chemicals, polymers for a variety
of end uses but mainly for paints and coatings, and
additives such as flame retardants, plasticizers and
antioxidants for plastics account for over half of
all the uses of these new chemicals.

Experience to date reveals the great majority of
PMNs to be submitted by large companies, those with
annual sales in excess of 100 million dollars. These
data alone do not permit either a conclusion that small
companies develop very few chemical products nor a
conclusion that the PMN requirements of TSCA have
severely hindered small chemical companies in their
new product innovation efforts. Reference was made
to a published study by the Chemical Specialties
Manufacturers Association (CSMA) which found that a
misperception by industry of PMN testing requirements
was a principal reason for the apparent decline in
introduction of new products by small ingredient
manufacturing firms.

Some confirmation of this cause was obtained by
EPA but more direct and valid information is needed
before definitive conclusions can be drawn. There is
no doubt, however, that as with any government regulation, increased cost is involved. Established
chemicals can more easily carry the burden whereas
new chemicals whose future is in doubt could prove to
be cost prohibitive in terms of regulatory compliance.
For this reason, the PMN program is under close
scrutiny by Vice President Bush's task force on regu-
atory relief. EPA is studying more cost-effective
means of compliance. Voluntary programs are being
examined, as well as elimination of unnecessary burdens
on various segments of industry.

The results of this examination of the PMN requirements of TSCA has prompted EPA to launch a wide-ranging program to reduce the regulatory burden imposed
on industry by this provision of the law and, con-
currently, in cooperation with industry, to educate
the small business segment with respect to the goals
and requirements of TSCA.

A joint EPA-industry program is proposed to make
it easier for the chemical industry to comply with the
PMN requirements of TSCA and hopefully encourage
more new product development activity. This program
includes a simplified reporting form promulgating
exemptions for those classes of new chemicals

expected to pose no unreasonable risk to health or
the environment, and several direct assistance programs
aimed at helping small chemical firms to comply with
TSCA. Joint EPA-industry efforts were proposed to
encourage industry to continue its new product
development activity through a better understanding
of the goals and regulatory requirements of TSCA.

With all of the introduction to this subject which you have
had today and after all of the excellent preceding presentations,
I am going to jump right into my subject and dispense with any
background briefing.

First, of all, I would like to clarify the term "new
chemicals." I am referring to the TSCA definition as those
chemicals not listed on the TSCA Chemical Substance Inventory and
maintained on a daily basis by the Office of Toxic Substances
within EPA. This is a list of all commercial chemicals - some
55,000 in all - produced in or imported into the United States
during the period of 1975 through 1979. My talk this afternoon
will not cover the thousands of formula changes in mixtures of
chemicals which occur almost daily as industry tries to meet
changing market demands.

As you heard this morning from several speakers, the premanufacture notification provision of TSCA has been in effect
since July 1, 1979 and since then EPA has received over 1,000
notices of intent to manufacture and introduce new chemicals into
U.S. commerce. This is the only complete and accurate record of
the development and commercialization of new chemicals ever compiled and, as such, is a repository of a wealth of information. A
major portion of it is classified by EPA as confidential business
information - CBI in our lingo - and is closely protected against
inadvertent disclosure.

Nevertheless, it is possible to analyze the information
supplied by industry on new chemicals and summarize it in a way
which does not breach CBI. This is what I have done in preparing
this paper and it is the work of many of my cohorts within the
Office of Toxic Substances. I intend to summarize the experience
of EPA in dealing with these notices including an analysis of the
classes and types of new chemicals, market areas, company size
and other data. From this I will draw some conclusions about the
impact of this requirement of TSCA on new product innovation and
will describe what EPA is doing about it.

On my first figure (Figure 1) I have plotted the number of PMNs
submitted per quarter and you can see the rapid rate of increase
since July 1, 1979. The rate of increase has slowed but there
doesn't appear to be any leveling off as yet. The total number
submitted during the first quarter of this year - ending today as
a matter of fact - will be about 210 but this includes a single
category of 30 or so chemicals.

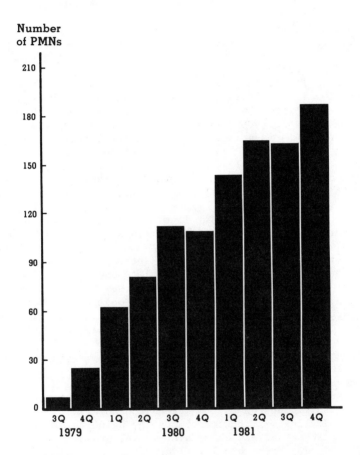

Figure 1.   Premanufacture notices (July 1, 1979–Dec. 31, 1981). The total number of PMNs submitted over the 10-quarter period was 1056.

On Table I is a list of the major end uses
for the new chemicals submitted up through the end of 1981. Intermediates in the manufacture of other chemicals, polymers for a
variety of end uses but mainly for paints and coatings, and additives such as flame retardants, plasticizers and antioxidants for
plastics account for over half of all the uses of these new
chemicals. These seven major categories in total represent
slightly over three fourths of all projected uses. One would
suspect that this pattern will change with market demand and
competitive developments and a year from now we might see intense
R&D activity in some other specific market areas culminate in the
introduction of a line of new chemical substances.

On Figure 2 I have shown the number of these
new chemicals which have actually been commercialized by the
manufacturer or importer as of March 1st of this year. About 30%
of the total submissions were commercialized through the end of
the 3rd quarter of last year. This seems like a very slow rate
of commercialization but the reasons for some companies not following through the development phase with a commercialization phase
are common-place in the industry. We made a spot telephone survey
of a number of these manufacturers and received the following
answers to our questions (Table II). The chemical industry would
say that these reasons are par for the course and undoubtedly
could add a few more.

We also came up with another interesting statistic in our
analysis of these data. On this table (Table III) I have summarized
the frequency of commercialization within certain time frames
after the expiration of the mandatory 90 days from receipt of the
PMN by EPA. As you can see, most companies want to produce and
sell their new chemical as soon as possible. We have not had time
to analyze these data further but it would be interesting to see
if there is any relationship between commercialization time and
type of chemical, end use or market, size of company, volume of
production, etc.

Now what have we learned about the companies which have
submitted these PMNs? Here, on Figure 3, I show the number of PMNs
submitted per company as a function of the number of companies
for the period from July 1, 1979 through the end of 1981. For
example, 61 companies each submitted one PMN, 36 companies
submitted two each, and so on. There were 25 companies which
submitted more than 10 PMNs each and I can add that the three
most prolific developers of new chemicals during this 2 1/2 year
period as judged by this yardstick each submitted 60 or more PMNs.
With time, this curve is tending to flatten out as more of these
companies submit additional PMNs and at a faster rate than new
companies enter the PMN area with their first submissions.

Another way of looking at these data is shown on
Table IV. For this period 186 different companies submitted a total of 1,056 PMNs. Here you also can see that 6 companies
representing only 3% of the total accounted for 289 or 28% of all

Table I. Major End Uses for PMN Chemicals

Intermediates

Polymers for paints and coatings

Additives for plastics

Additives for lubricants and
   cutting fluids

Dyes, inks and related products

Polymers for adhesives

Photo products

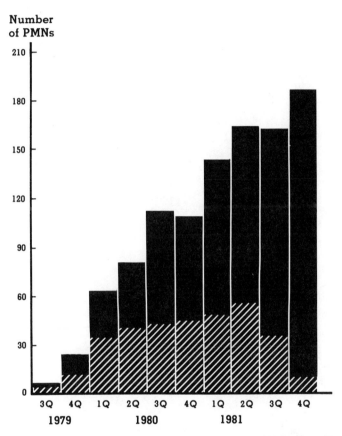

Figure 2.   Premanufacture notices (July 1, 1979–Dec. 31, 1981). Key: ■, *total PMNs; and* ▨, *PMNs for which commercial production has commenced as of March 1, 1981.*

Table II. Reasons for Not Commercializing New Chemical
Substances

o cost of raw materials increased to
   the point where the final product
   was priced out of the market

o on scale-up, the product didn't meet
   performance goals

o awaiting patent clearance before
   commercial sales begin

o customer solved his problem and didn't
   need the new product

o other higher priorities for capital
   funds forced a delay of this project

Table III. Commercialization Times for New Chemicals

Time Frame	Number of New* Chemicals Commercialized	% of Total
within 1 month	126	40
1 to 3 months	90	29
3 to 6 months	46	15
over 6 months	51	16
	Total = 312	100

*As of March 1, 1982

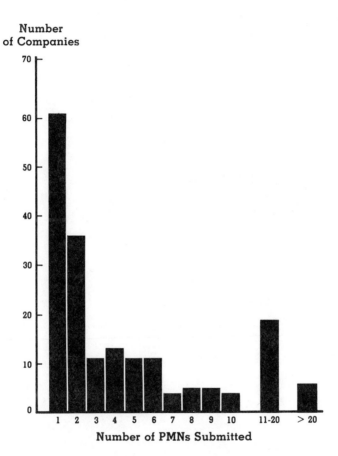

Figure 3.   Number of PMNs submitted (total = 1056) vs. number of companies (total = 186) from July 1, 1979 to Dec. 31, 1981.

Table IV. Premanufacture Notices
(7/1/79 to 12/3/81)

	PMNs		Companies	
No. per Company	Total No.	% of Total	Number	% of Total
1	61	6	61	33
2 to 5	212	20	71	38
6 to 10	219	20	29	16
11 to 20	275	26	19	10
over 20	289	28	6	3
	1,056	100	186	100

PMNs. Fewer than 15% of the companies submitted over one half of all PMNs.

We attempted to relate the number of PMNs submitted to the size of the submitting company as judged by dollar sales. Our data here are not absolutely firm but I think we are reasonably accurate in the tabulation shown on Table V. We used the total annual sales of a company if it was a wholly-owned subsidiary of a larger company including foreign-owned multinationals. As you might expect our only difficulty in finding these annual sales figures was in the industry segment composed of small, privately-held concerns. For this period the great majority of new chemical substances was developed by the larger manufacturing firms.

So these are the facts. Now what conclusions can one draw. I for one believe these data alone do not permit either of the following conclusions: (1) that small chemical companies do very little development of new chemical substances, or (2) that the PMN requirement of TSCA has seriously hindered small chemical companies in their development of new chemical substances. Additional information is needed before a valid choice can be made between thes two alternative conclusions or even if there is some truth in both.

Jack Yost has just given you a first hand report on how TSCA has affected his company's operations. Also, CSMA recently surveyed their membership on this specific subject, and found that there were two principal reasons for an apparent recent decline in development of new products by small ingredient manufacturing firms. One was the wide misperception in the industry that health and safety testing is required before EPA will process a PMN. Neither the law nor EPA regualtions requires a firm to perform any testing prior to filing or having EPA process a PMN. It is only in those rare cases in which EPA believes there is evidence a new chemical substance could present an unreasonable risk that testing can be required under Section 5(e) of TSCA and commercial production can be delayed. Because a 5(e) order cannot be issued without cause, EPA has taken this action on fewer than 1% of the PMNs submitted.

And, in addition, in anticipation of a possible 5(e) order because of EPA concerns about potentially harmful health or environmental effects, 4 companies have withdrawn their PMNs to develop additional data or discontinue commercialization efforts.

Let me hasten to add here that EPA is not encouraging companies to submit PMNs devoid of data. Quite the contrary. The quality of our risk assessment of a new chemical is directly related to quality and quantity of the health and environmental information we receive from the submitter or are able to obtain from the literature and all under the pressure of a 90 day time limit. Industry understands our position and is responding very well to meet our needs. We are receiving more pertinent data on new chemicals and especially from those companies coming in with

Table V.   Size of Companies Submitting PMNs
(7/1/79 to 12/3/81)

Company Size* ($ sales)	Number of PMNs	% of Total
Less than $10 MM	19	2
$10 MM to $100 MM	123	12
$100 MM to $500 MM	164	15
Over $500 MM	750	71
	1,056	100

*includes parent company

their second, third and subsequent PMNs. Telephone response to
our questions during the review period is supportive and gratify-
ing.

A second reason uncovered in the CSMA survey is the substantial cost impact associated with the commercial practice of
requiring completion of the PMN process before a customer will
accept a new chemical substance for evaluation and product testing. This practice aggravates the first reason and the combination is apparently causing a significant reduction in new
product innovation among smaller specialty chemical firms - those
with annual sales in excess of $200 million - are reducing their
new product development efforts. And, certainly judging by the
past 2 1/2 years of PMN experience, we in EPA corroborate the
CSMA findings even though there is no accurate record of
performance in pre-TSCA days.

In my own direct contacts with industry, I have heard many
times that small firms just don't have the personnel to assemble
the pertinent data and fill out the PMN form for a new chemical.
So they studiously try to solve their customer problems by using
chemicals already on the TSCA Inventory.

Before I pursue this problem any further, let me point out
the plus side of this provision of TSCA adding to what was discussed this morning by Ham Hurst. Today, industry leaders
generally agree that TSCA is having a positive and beneficial
effect on their attitude and behavior toward their manufacture,
distribution, processing, use and disposal of chemicals. As
Dan Harlow just reported, management is taking a more responsible
look at the possible effects of their products on the health of
workers and consumers and on the environment. The attitude has
changed and that is a prerequisite for achieving the goals of
TSCA.

We have no way of knowing just how many really harmful
chemicals were kept off the U.S. market since the PMN requirement
of TSCA went into effect. We are not even sure that some of the
1,000 plus new chemicals which have passed through the PMN process
will not prove 20 or 30 years from how to be serious carcinogens
or mutagens. What we do know, however, is that these 1,000 new
chemicals have been subjected to a rigorous examination with
respect to potential risks to society by both the business
community and by the Federal Government.

The Office of Toxic Substances has assembled a team of multi-
disciplined scientists to review each of these PMNs and assess
the potential risks to human health and the environment posed by
commercial manufacture and sale. These assessments are based upon
limited firm data on the specific chemical, comparison with
structurally similar chemicals of known toxicity, plus estimates
of exposure from calculations of the potential number of people
involved in manufacturing and processing operations and in
consumer use. Most PMNs contain elementary data on physical and
chemical properties and obvious acute health effect such as skin

and eye irritation. We are beginning to receive more information
on possible chronic health and even environmental effects particularly from those companies which have submitted several PMNs.
Some data on mutagenicity using an Ames Test and indications of
persistence in the environment using octanol-water partition
coefficients are included.

EPA plans to follow up on selected new chemicals during the
commercialization phase. This program will be focused on those
chemicals of concern as well as those for which there is uncertainty concerning toxic effects. Unrestricted commercialization
could lead to substantial increases in exposure so that it may be
necessary to reassess the risk through additional testing. TSCA
grants EPA the authority to do so under Sections 5(a)(2) and 8(a)
but voluntary action will be sought where appropriate.

I think it is very safe to say that this risk assessment
process for new chemicals was never applied so thoroughly in pre-
TSCA days and in this respect TSCA is meeting one of its major
goals.

But as with any government regulation, there is an accompanying cost to society to achieve these benefits and the Congress was
particularly concerned with the potential inhibiting effect of
TSCA on innovation in the chemical industry. Regulatory costs are
more easily borne by established commercial chemicals than by
speculative new chemicals whose commercial future is in doubt. On
the other hand, Congress recognized that any preventive regulatory
action of TSCA with respect to hazardous new chemicals entering
the marketplace can be achieved with less cost to industry in
terms of loss of jobs, profit and capital investment.

Under the new administration, this section of TSCA has come
under the scrutiny of Vice President Bush's task force on
regulatory relief and the Office of Toxic Substances is placing a
high priority on efforts to develop more cost-effective means for
achieving industry compliance with OTS policies. In addition to
fostering voluntary actions by industry wherever possible in lieu
of formal rules, these effects include elimination of unnecessary
burdens on industry in complying with mandated TSCA requirements.

Beginning with PMN forms, our current plans are to require
only that information clearly spelled out in the statute with all
other information being optional. Our experience with submitters
to date is that informal requests for additional information have
generally proved adequate for our risk assessment needs so we will
continue to rely on this approach. This should clear up the uncertainty on the PMN review process and reduce the burden on
industry.

The Office of Toxic Substances is devoting substantial
resources to issuing exemptions to PMN requirements which should
still further reduce the regulatory burden. Naturally, these
exemptions will cover only those new chemicals which are expected
to pose no unreasonable risk to health or the environment. As you
heard this morning from David Zoll of CMA, we are approaching this

Table VI.   Program to Minimize Negative Impact of TSCA
on New Product Innovation by the Chemical Industry

EPA

1- Promulgate concise and simple rules for PMN
information requirements

2- Promulgate rules exempting from PMN requirements
certain classes of chemicals considered to pose
no unreasonable risks to health and environment, e.g.

     o high MW polymers
     o site-limited intermediates
     o low volume chemicals

3- Provide and publicize information consulting services
to industry for preparing PMNs

Industry

1- Take advantage of EPA's flexible treatment of PMN
informational needs for risk assessment

2- Take greater advantage of EPA assistance in planning
and preparing PMNs

Joint EPA and Industry

1- Conduct a wide-ranging campaign to educate small
business firms about PMN requirements and
available assistance services.

regulatory relief in cooperation with industry and public interest groups on several specific proposals made by industry. These include certain high molecular weight polymers, site - limited or captive intermediates, and some relatively low production volume cut-off. Some way of reducing the mandatory 90 days reporting requirement is also being explored which could materially benefit the small chemical concern primarily in toll business and customer problem-solving.

So my message to the chemical industry broadly and in particular to the segment of small businesses is to not let the PMN requirements limit your creative spirit in the development and commercialization of new chemicals. The Office of Toxic Substances in EPA stands ready to assist you at no cost in filling out the forms and distinguishing clearly what is minimally essential for risk assessment purposes from what is optional. We have trained consultants in the northeastern and mid-west sections of the country ready to travel to your location and advise you in a confidential way on PMNs. Our staff in Washington can be immensely helpful and you should feel free to discuss your specific situation with them. And, to top it off, we have one recently-retired experienced chemcial industry man concentrating his entire effort to assist the small business man in any TSCA-related matter. His name is Dr. Bob Toomey; give him a call on the Industry Assistance Office toll-free line.

I have discussed this concern of EPA on the possible negative impact of TSCA on product innovation with many people and groups representing different segments of the chemical industry. You heard Carl Umland this morning on this same subject and there is agreement that something can and should be done. From the government perspective, the following program makes sense and we intend to pursue it aggressively (Table VI).

Simplified notification rules are a must and this is a high priority item. I mentioned earlier our efforts to develop helpful exemptions to PMN requirements under section 5(h)4 of the Act and several proposals are being readied for publication and comment. And this action is important for EPA to provide direct assistance to small business concerns in preparing PMNs.

We would like to encourage the industry to submit PMNs and have companies work closely with us to assess potential risks. This can be done without great expense to manufacturers. And, finally, a well-organized and continuing joint industry-EPA program to help small business understand TSCA and ease the reporting burden would go a long way to encourage the development of new chemical products by the chemical industry.

RECEIVED January 17, 1983

Future for Innovation

CARL W. UMLAND

Exxon Chemical Americas, Houston, TX 77001

Chemicals have traditionally stimulated technological
progress and contributed substantially to our quality of life.
The U.S. chemical industry has been highly innovative in the
past. The legislative history of the Toxic Substances Control
Act (TSCA) demonstrates Congressional concern that the economic
impact of the Premanufacture Notification (PMN) process could
unduly interfere with the innovative capacity of a dynamic
contributor to the economy. At the same time, the Act was to
assure that innovation and commerce in chemical substances do
not present an unreasonable risk of injury to health or the
environment. After five years of TSCA and nearly three years of
PMN activity, some evidence based on actual PMN experience has
emerged which suggests substantial disruption of new chemical
development and introduction. This appears to be largely a
consequence of higher costs and uncertainty engendered by the
PMN process at an economically vulnerable point in the life
cycle of innovative products. Reassessment of the process
seeking a more appropriate balance between opportunity for
economic viability and protection from unreasonable risk for
innovative chemicals is indicated. An intelligent and pragmatic
application of TSCA exemption authority to well-defined low risk
situations offers significant hope for improvement in the
outlook for chemical innovation within the spirit of TSCA but
well short of laissez faire "business as usual".

The chemical industry is linked to virtually every segment
of the American economy as shown in Figure 1. Chemicals are
used by other industries as feedstocks, cleaners, additives, and
processing aids for a wide range of products and industrial
processes. The chemical industry also provides consumer
products directly. These range from soaps and detergents to ink
and paint. The effects of chemical innovations, then, are felt
far beyond the chemical industry.

0097-6156/83/0213-0023$06.00/0
© 1983 American Chemical Society

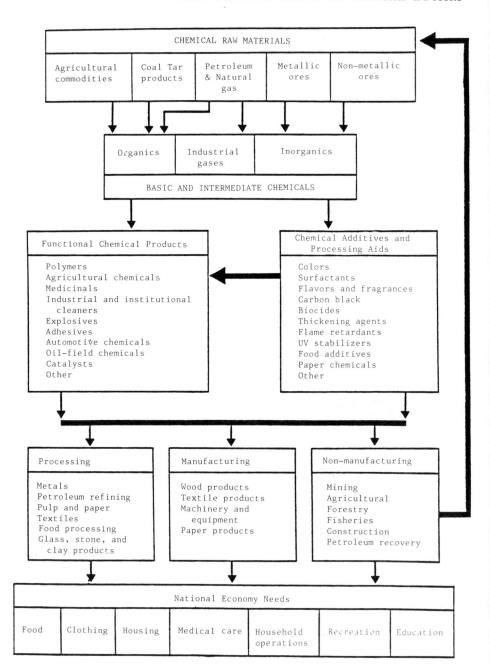

Figure 1.   Materials and product flow of chemical industry. (Reproduced with permission from Ref. 13. Copyright 1980, C. H. Kline & Co.)

Even individual segments of the chemical industry have a very large economic importance. As one example, an Arthur D. Little study has estimated that 23 percent of all business sales, 16 percent of all capital investment and 19 percent of total non-government related jobs are dependent on the production of petrochemicals (1). According to this study, 35 to 45 percent of United States business activity is directly or indirectly affected by the American petrochemical industry.

The Nature and Process of Chemicals Innovation

Industry's concern over the potentially stifling impact of TSCA's PMN process on innovation derives from the nature of new chemicals commercial development. Technological progress usually consists of a series of small innovations which must survive the test of the open market. Increased costs, especially in the initial stages, can easily put an innovation at a fatal disadvantage in competing with established products. PMN costs could stop the process by which small, individual innovations contribute to significant technological changes.

New chemicals must have time to prove themselves in the marketplace. Often, the true value of an innovation is not known for many years after its introduction. The initial use of a new product is industry's opportunity to appreciate the chemical's properties and explore its potential. A new chemical often initially produced in small quantities, may become a large scale venture after many years with significant and occasionally unanticipated applications.

PMN costs could choke innovation at its most vulnerable stage. A new product must have "breathing room" in which to test its potential for growth. Decisions to discontinue research and development on many chemicals because of higher costs could reduce major technological breakthroughs in the future.

Innovation particularly in small volume chemicals often depends on a quick reaction to market opportunities and low costs. Six major characteristics of the chemical innovation process for small volume chemicals lead to this conclusion.

Small volume chemicals are often brought into the market quickly. Chemical specialties especially are created and marketed to meet an immediate need. A delay in a chemical's introduction could mean that the user would seek materials elsewhere and certain market opportunities would be lost.

Some chemicals sold in small volumes cannot withstand a lengthy research and development process. Large R&D costs would be spread over too low a sales volume to allow the product to compete with existing chemicals. Only those which are obviously essential or obviously provide significant technological advance can support high initial costs.

Delays in developing a new product significantly increase development costs. In reviewing the PMN requirements, the Council on Wage and Price Stability using a 10-percent discount rate, observed that the cost to the firm of a six-month delay was about five percent of the R&D investment (2).

Small volume chemicals are an extremely important part of chemical substances. Production data reported to EPA suggest this strongly. In Congressional testimony, then EPA Assistant Administrator Steven Jellinek reported that about 70 percent of the chemicals in the EPA inventory are produced in quantities under 100,000 pounds per year, 50 percent are under 10,000 pounds and 30 percent are under 1,000 pounds (3). A further example is the testimony of the Reilly Tar and Chemical Corporation which reported that 51 percent of its pilot plant (newly introduced) products had annual sales volumes under 50 kilograms while only 17 percent had sales over 1,000 kilograms (4).

The sales of many new chemicals remain small for several years after introduction and about half of new chemicals are discontinued because they are not commercially viable. The chemical group of the Ansul Company provided a useful example in its November 1979 comments on Premanufacture Notification of how the sales volume of one of its typical chemical products grew from 1,000 pounds to only 10,000 pounds in five years (5). Changes in demand, breakthroughs by competing companies, fluctuations in the price or availability of raw materials and faulty original estimates can all cause a product to fail. Such market performance dramatizes the uncertainty of marketing new products. Indeed, experience has shown that about half of new commercial ventures fail.

Important uses for new chemical substances have often been discovered many years after their commercial introduction. Some of today's most important chemicals, such as resins or plastics were commercially unimportant when they were first introduced. The original uses of new chemicals are slowly supplanted by new applications which increase their production. Therefore, a new chemical must remain commercially viable long enough for new uses to be discovered.

Teflon   Fluorocarbon Resin, for instance was discovered in 1938. The substance was produced on a small scale throughout World War II for defense applications. It became commercially available in 1948, but did not gain public recognition until 1961 when it was used as a lining for nonstick frying pans. Today, Teflon   coats chemical process equipment, insulates electrical equipment, coats the blades of tools, protects buildings from earthquakes and is used as a material in

automobiles, pianos, missiles and spacecraft. Most of these
uses were unknown when Teflon was invented.
Important uses for new chemical substances have often been
discovered many years after their commercial introduction. Some
of today's most important chemicals, such as resins or plastics
were commercially unimportant when they were first introduced.
The original uses of new chemicals are slowly supplanted by new
applications which increase their production. Therefore, a new
chemical must remain commercially viable long enough for new
uses to be discovered.
Teflon Fluorocarbon Resin, for instance was discovered in
1938. The substance was produced on a small scale throughout
World War II for defense applications. It became commercially
available in 1948, but did not gain public recognition until
1961 when it was used as a lining for nonstick frying pans.
Today, Teflon coats chemical process equipment, insulates
electrical equipment, coats the blades of tools, protects
buildings from earthquakes and is used as a material in
automobiles, pianos, missiles and spacecraft. Most of these
uses were unknown when Teflon was invented.

The development costs of most new chemicals cannot be
readily passed on to consumers. At the time of their
introduction, most new chemicals represent a minor improvement
over existing materials. Customers must believe that the
product is less expensive or more efficient than the existing
product. Only then will the new product have an opportunity to
expand its role in the economy.
Even small increases in price could eliminate the advantage
that a new product might enjoy over its already established
rivals. Higher prices coupled with the uncertain performance of
a new product and lack of customer familiarity with the chemical
may deter customers from buying the new material. Thus, an
increase in cost generally will not be reflected in a higher
price, but in a decision by the manufacturer to forego
production.
One major chemical specialty company, for instance, uses a
10-percent product improvement as its criterion for determining
whether to market a new product. That is, the product must have
a 10-percent improvement in performance at the same price or the
same performance at a 10-percent cost reduction. From its
experience, the company knows that a new product will not be
accepted in this company's markets without such an improvement.
For this company, a typical chemical's cost is about 50
cents per pound. The firm generally assumes no more than a
three-year marketing lifetime for its type of new chemicals.
If one then assumes that the corporation has to recover its
PMN costs during the lifetime of a chemical, the percentage
increase in cost can be calculated for different PMN costs as
shown in Table I.

Table I. Product Cost Increase

Annual Production	PMN Filing Cost			
	$200	$2000	$7000	$15000
2500 Lbs.	5%	50%	187%	400%
25000 Lbs.	0.5%	5%	19%	40%
100000 Lbs.	0.1%	1%	5%	10%

One can readily see the intense volume and PMN cost sensitivity over and above this company's own decision criteria of a 10% improvement over existing products before consideration for marketing. Thus, the hurdle level created by PMN costs can be far greater than a company's original decision criterion and the chances for small volume substances being brought forward are reduced. Obviously, with longer product life the effect would be less dramatic but no less real.

The cost of developing a new chemical will not be spread over the manufacturer's product line. As a general rule, no company will intentionally begin a venture unless it believes that the venture will recover its investment and make a profit. If a manufacturer does not believe that the product will generate an acceptable return on investment, the company will invest its funds elsewhere. Any other approach would be uneconomic and would eventually harm the company.

Product failures, on the other hand, become a fixed cost to the company and must be recovered through higher prices for other products if the firm is to remain viable. Thus, there is an increased incentive to avoid the economic uncertainty of innovation.

According to the Council on Wage and Price stability "Studies of business innovation suggest that over the long term companies treat their research and development budgets like other investments and adjust R&D expenditures so that the return is comparable to that earned on other corporate investments", (2).

TSCA Impact on Innovation

We now have nearly three years of PMN experience. An analysis of both the number and the character of new substances introduction as represented by PMN filings suggests that there has indeed been a substantial negative impact on chemicals innovation. This observation is bolstered by a broad variety of studies and surveys which are highlighted in what follows. Limited information exists on the rate of new product

introductions before TSCA. Arthur D. Little, Inc. estimated
that 1,000 commercial new chemicals were introduced each year,
(6). An earlier study by Foster D. Snell, based on expert
industry opinion, found that 2,220 successful new substances
were introduced each year between 1969 and 1974 out of over
5,000 offered for customer evaluation. National Economic
Research Associates, Inc. (NERA) estimated that about 1,700 new
chemical substances were sold commercially each year (based on a
small sample and accurate only within a broad range but
consistent in order of magnitude) (7). Nevertheless, the
introduction of new commercial chemical substances as measured
by PMNs submitted apparently has fallen from between 1,000 and
2,200 annually to somewhere around 600-700 in the latest
12-months. Without trying to be precise, it can be seen that
there has been an apparent drop in new substance introduction in
the order of 35-70 percent. This is even more sobering when it
is realized that about one-third of PMN chemicals have actually
entered commerce based on a more detailed presentation of the
PMN statistics by EPA's D. G. Bannerman.

However, for the purposes of this discussion, one can
observe that of the more than 1,000 PMNs submitted, over 90% are
from large companies. Furthermore, estimated first year
production volumes have steadily moved away from low volume
chemicals. Arthur D. Little, Inc. estimated that prior to the
PMN requirement about 70 percent of commercial new chemicals
were produced in quantities under 1,000 pounds per year and all
R&D chemicals were below this level (6). When the PMN
requirements went into effect, however, that proportion fell to
33 percent almost immediately and has since declined to only 11
percent, based on a look at the first 723 PMNs. This can also
be compared with the 30 percent of commercial chemicals on the
TSCA inventory with an annual production volume of less than
1,000 pounds.

Figure 2 illustrates the decline in the distribution of low
volume chemicals between 1978 and October 1981 rather
dramatically.

The samples show that PMN requirements have changed the way
that companies do business and that there are fewer chemicals
with small production volumes being introduced. The decline in
the production of new, small volume chemicals suggests that
chemical companies are now concentrating on products that have
ready markets.

The smaller chemical specialty manufacturers have been hurt
the most. A recent study commissioned by the Chemical Specialty
Manufacturers Association indicates that while new substance
development fell 26 percent among chemical specialty
manufacturers, 98 percent of the decline was concentrated among
ingredient suppliers with less than $100 million of annual
sales (8). Although the general economic climate may have

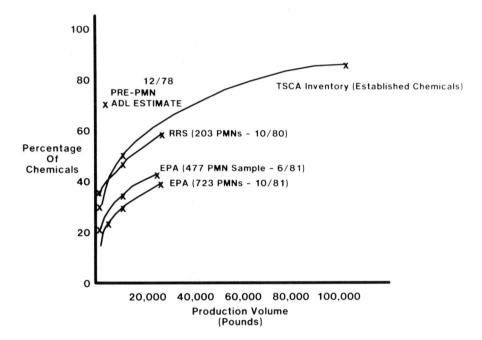

Figure 2. Distribution of established chemicals and first-year chemical production under PMNs.

contributed to some of this decline, it does not explain why such a disproportionate share of the drop should settle upon small companies.

The drop in innovation was most pronounced in the more financially risky types of chemicals. Innovations aimed at the general chemical market without defined applications declined 38 percent while chemical innovations manufactured at specific customer request fell only 14 percent.

The reduction in innovation because of financial risk was predictable. From studies for EPA by Arthur D. Little, one would have estimated a 50-percent reduction associated with a $12,000 PMN cost (6). The significance of $12,000 derives from a study by the Regulatory Research Service (RRS) (9).

RRS assessed the cost of completing the PMN form by examining the costs that companies actually incurred in performing this task. RRS sent a questionnaire to all companies known to have filed a PMN form. In addition, RRS examined EPA's public files on the PMNs it has received. Average costs are shown in Table II.

Table II. PMN Filing Costs

(PER RRS)

	DOLLARS
Mean Total Filing Cost	7500
Economic Burden of Intermediates	2300
Economic Burden of Adverse PMNs	2200
	12000

Intermediates don't often reach the marketplace but do require a PMN under TSCA. Similarly, many chemicals apparently had PMNs filed to be ready for manufacture/marketing but then failed to make it through the business decision process for other reasons.

The $12,000-cost of filing a PMN is still a conservative figure. The additional cost of customers requiring a PMN on developmental material would add another $10,000 to the cost of a PMN. Such costs are beginning to appear and are becoming significant in some segments of the chemical industry, but are not included in this cost calculation. In addition, the RRS figure does not include a provision for commercial ventures which fail after the chemical enters the marketplace.

The Future for Innovation

The situation described up to this point is distressing and needful of remedy. However, we've only addressed the economic part of the equation. We also have to consider what kind of

risk to health or the environment has been represented by the
PMNs submitted to date.

Fortunately, that experience has been assessed and found to
offer a reasonable basis for a change in course. Recent
statements by EPA officials point the way to a potential for
relief.

Mr. Don Clay, Director of the Office of Toxic Substances,
discussed the premanufacture review procedures and experience
with PMNs to date at a meeting of the Organization for Economic
Cooperation on Development (OECD) Chemicals Forum in December,
(10). He noted that EPA's chemistry, toxicology, and exposure
assessment teams normally complete their preliminary evaluation
within a week of receipt of a PMN, and, that preliminary
assessment eliminates about 50 percent of the substances as
chemicals of low concern. They then proceed to structure
activity analysis and reasonable worst case assumptions to
assess unreasonable risk or the need for more data.

The results of that process were commented on by Dr. John
Todhunter, EPA's Assistant Administrator for the Office of
Pesticides and Toxic Substances (OPTS) in a speech delivered in
Rome to representatives of some of our European trading
partners (11). He was commenting on the small amount of
toxicity data submitted with many of the PMNs submitted to
date. He pointed out that EPA's experience shows that to be due
largely to the inherently low hazard potential of the bulk of
the substances submitted for PMN review. His evidence was the
fact of no imminent hazard actions (Section 5(f)) and inadequate
information actions (Section 5(e)) on only 9 chemicals out of
over 1,000 PMNs submitted. He further noted that in 60 cases,
industry had volunteered more data, reduced exposures, or
withdrawn PMNs. Dr. Todhunter's conclusion from all this was
that industry is doing an effective job of screening substances
before submitting PMNs.

The Chemical Manufacturers Association (CMA) had reached
similar conclusions about a year ago and filed a petition with
EPA suggesting that there was a strong case to be made for
exemptive relief under Section 5 (h)(4) of TSCA for many
polymers, site-limited intermediates, and chemicals produced in
volumes of less than 25,000 pounds per year. (An examination of
the effect of PMN costs at various prices and levels of
production reveals that the PMN cost, as a percentage of total
cost per pound of product, generally rises most rapidly as
output falls below 25,000 pounds. See the Appendix for a more
detailed discussion.) It is CMA's firm belief that exemption
could be granted in terms that would operate to assure "no
unreasonable risk" to the public in terms of either health or
the environment.

Subsequent open discussions with EPA have proven helpful to
an understanding of how such an exemption might be granted with

adequate safeguards to operate effectively. As a result, there
is a ray of hope that EPA will soon propose an exemption rule
for public comment as a first step to a sound and rational
correction of the undue inhibition of chemicals innovation which
has occurred over the past three years.

Impact of an Exemption

Most polymers are inherently non-toxic and can be
sufficiently defined to present no unreasonable risk.
Site-limited intermediates have limited exposure potential by
definition which together with chronic hazard control language
will present no unreasonable risk but will result in real
economic savings.
A 25,000-pound per year production rate is not a
demarcation between large and small ventures. Such a point
would be at a far higher scale of production. Instead, the
proposed 25,000-pound exemption represents an economically-
justified and virtually risk-free means of aiding innovation in
the chemical industry, particularly when coupled with appropriate chronic hazard control language as for intermediates.
In fact, chemicals produced in quantities of 25,000 pounds
or less per year comprise a negligible proportion of chemical
output. Chemicals listed in the TSCA inventory had a combined
production of nearly 4.1 trillion pounds in 1980. Chemicals
that have production volumes under 100,000 pounds per year
contribute only 0.006 percent to the total (12). A 25,000-
pound exemption, then, would free an insignificant proportion of
chemical production from reporting requirements.

Conclusion

With such exemptions, innovative chemicals can be produced
to explore the commercial market and to test product viability.
Knowledge of their economics and potential for different
applications can be expanded. These exemptions will lessen the
cost of market failures. Successful chemicals will not have to
bear the additional PMN cost of unsuccessful market tests. As
new substances demonstrate their value to society and find a
secure place in the market, they can then begin to absorb some
of the costs associated with the PMN requirements. There would
indeed be a chance for regaining much of the strength of an
innovative chemical industry with all of its attendant benefits
to society within the context of TSCA and the intent of
Congress' expression of the will of the American public when it
it was passed with the support of the chemical industry in 1976.

Literature Cited

1. Arthur D. Little, Inc., "The Petrochemical Industry and the U.S. Economy"(1978).
2. Council on Wage and Price Stability, "Premanufacture Notification under the Toxic Substances Control Act " March 13, 1981.
3. Testimony of Steven Jellinek before the Subcommittee on Consumer Protection and Finance of the House Committee on Interstate and Foreign Commerce, March 20, 1979.
4. Reilly Tar & Chemical Corporation, "Proposed Premanufacture Notification Requirements and Review Procedures. Comments for the U.S. Environmental Protection Agency", March 21, 1979.
5. Ansul Company, "Comments on EPA Reproposal of Premanufacture Notification Form and Provision of Rules", November 29, 1979.
6. Arthur D. Little, Inc., "Estimated Costs for Preparation and Submission of Reproposed Premanufacture Notification Form", September 1979.
7. National Economic Research Associate, Inc. Paper by Lewis J. Perl, "The Impact of Toxic Regulations on Industry", October 1981.
8. Chemical Specialties Manufacturers Association, "Impact of the Toxic Substances Control Act on Innovation in the Chemical Specialties Manufacturing Industry", January 1982.
9. Heiden, Edward J. and Alan R. Pittaway, "A Critique of the EPA Economic Impact Analysis of Proposed Section 5 Notice Requirements", Regulatory Research Service, (1981)
10. Don R. Clay, Director OTS-EPA, "The Premanufacture Review of New Chemicals Under the Toxic Substances Control Act", December 1981.
11. John A. Todhunter, Assistant Administrator EPA-OPTS, "Role of Administration and Industry in Implementing Hazard Evaluation Within a Regulatory Framework", December 1981.
12. Blair, Etcyl H., "A Framework of Consideration for Setting Priorities for the Testing of Chemical Substances," Workshop on the Control of Existing Chemicals, Berlin, Germany, June, 1981.
13. Curry, S.; Rich, S. "The Kline Guide to the Chemical Industry"; C.H. Kline & Co.: Fairfield, New Jersey, 1980; 4th Edition, p. 2.

APPENDIX

Rationale for Exempting Substances When

25,000 Pounds a Year or Less are Produced

Precedents for a 25,000-Pound Exemption

EPA has used a 25,000-pound exemption in its regulations governing recordkeeping in the management of hazardous waste under the Resource Conservation and Recovery Act (RCRA). The regulations list particular wastes as "hazardous" and establish standards that the generators of such wastes must observe.

The regulations provide a recordkeeping exemption for hazardous wastes created by a "small quantity generator". Such a generator is defined as one who creates less than 1,000 kilograms of hazardous waste per month. Since 1,000 kilograms equals approximately 2,200 pounds, the regulation effectively reduces the recordkeeping burden of firms that generate hazardous wastes in quantities of 25,000 pounds per year or less. If 25,000 pounds is too small a quantity of hazardous waste to bear careful scrutiny under RCRA, a 25,000-pound PMN exemption for low-toxicity chemical production is surely justified.

Moreover, the EPA Office of Pesticides and Toxic Substances has grouped pesticide active ingredients into three different categories based on production and exposure potential. One of EPA's three categories consists of "low" production pesticides. The Agency defined this category as an annual production volume of 25,000 pounds or less.

EPA's use of a 25,000-pound exemption (actually a 26,400-pound exemption) for hazardous waste recordkeeping and its definition of 25,000 pounds as low-level pesticide production indicate that EPA believes this volume of chemicals does not present a significant risk. A relatively safe substance produced in a volume of 25,000 pounds or less per year, then, would pose virtually no risk. Such an exemption from the PMN requirements would be reasonable.

Economic Burden of the PMN Requirement on Production Costs

A 25,000-pound annual production volume exemption is also a rational point at which to establish an exemption economically. Such an exemption would sharply reduce the economic impact of PMN requirements while protecting the public from unreasonable risk.

Presently, the cost of filling out a PMN form adds significantly to the unit cost of producing a small volume

substance. As previously discussed, these costs frequently
cannot be passed on to customers because of competition from
existing products. At low levels of output, the additional cost
per pound of substance rises dramatically.

Figure A shows the impact of $12,000 PMN-cost on the per
pound cost of producing a substance at low levels of output. A
substance which would normally cost 25 cents per pound
experiences a 20 percent cost increase at a production rate of
60,000 pounds per year. Costs rise by 32 percent if only 40,000
pounds per year are produced annually. Below 25,000 pounds of
output, the percentage increase in cost per pound begins to rise
almost vertically on the scale.

The impact on cost at 50 cents a pound is similar. The per
pound cost increases by 16 percent at 40,000 pounds of output
per year, 26 percent at 25,000 pounds per year, and 64 percent
at 10,000 pounds per year. Again, the cost impact becomes
increasingly severe at the lowest levels of production.

The effect is much the same at costs of $1 and $2 per
pound. The cost increase becomes increasingly dramatic as the
production level decreases. As Figure A shows, however, the
PMN-cost impact begins to weigh most heavily on each pound of a
substance at a level between 20,000 pounds per year and 60,000
pounds per year, depending on the cost of the product. At 25
cents per pound, costs begin to rise rapidly under 60,000 pounds
of output, while at $2 per pound the most rapid cost increases
occur under 20,000 pounds of production.

Many commercial chemicals sell for under $1 per pound. At
$1 per pound costs begin to escalate most rapidly under 25,000
pounds of output. An exemption, then, should not be set at less
than 25,000 pounds per year to minimize the cost of PMN forms on
innovative chemical products.

Even the spot price of most chemicals is under $2 per
pound. Spot prices on September 28, 1981, as published by the
Chemical Marketing Reporter, indicated that of 2,684 chemicals
sold by the pound or kilogram, 1,516 or 56 percent were priced
below $2 per pound. About 39 percent were priced below $1 per
pound, 19 percent were below 50 cents per pound and 4 percent
were below 25 cents per pound. Such prices tend to be higher
than the long-term contract prices under which chemicals are
usually sold. Thus, the curves in Figure A show the applicable
range of costs and output.

A lower exemption, 1,000 kilograms for instance, would
create an unnecessary impediment to innovation. New chemicals
would face an almost insurmountable cost barrier to expanded
production. The barrier would be much lower with a 25,000-pound
per year exemption. Innovation is not just the discovery and
testing of new products, but also those products' acceptance on
a commercial scale and at a competitive price. The impact of
PMNs on the cost of chemical substances was calculated using a

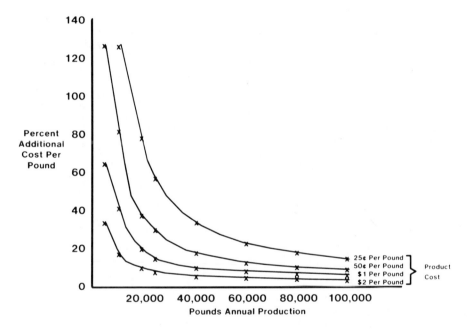

Figure A. Additional cost of PMN requirement as a function of annual production rate.

$12,000-cost for each PMN. This cost appears to be an accurate estimate because the Regulatory Research Service derived it from actual company experience in filing PMNs.

Companies were assumed to recover the PMN costs over five years. Most new products have very short lifetimes. Five years is reasonable.

The additional income required each year to recover the entire cost was derived by calculating the cash flow necessary to bring the net present value of PMN costs up to zero. A 10-percent discount rate was used in all calculations. The resulting annual cash flow was then divided by the pounds of output per year. Percentage changes were used to illustrate the relative burden of PMN costs.

Different assumptions yield similar results. Figure B illustrates the impact of assuming a $7,500 PMN-cost, a 15-percent discount rate, and either a three-year or ten-year payback period. The chemical cost is assumed to be $1 per pound.

Again, PMN costs begin to weigh heavily between 20,000 and 40,000 pounds of annual output. Under 20,000 or 25,000 pounds, the cost burden rises dramatically.

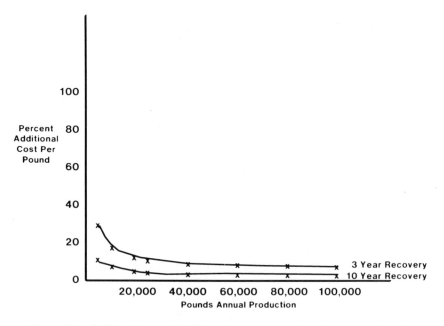

Figure B.   Additional cost of PMN requirement as a function of annual produc-tion rate using the following assumptions: base product cost, $1/lb; PMN cost, $7500; and discount rate, 15%.

RECEIVED September 1, 1982

Harmonizing the Regulation of New Chemicals in the United States and in the European Economic Community

BLAKE A. BILES

Jones, Day, Reavis, & Pogue, Washington, DC 20006

In the late 1970's, both the United States and the European
Economic Community (EEC) enacted laws -- TSCA and the Sixth
Amendment respectively -- that require companies to notify the
government before they commercialize new chemicals. Although
these notification programs are similar in many respects, their
enactment and initial implementation(2) have highlighted significant inconsistencies and, in a few cases, direct conflicts. As
a result, governments and industry on both sides of the Atlantic
have focused considerable attention upon the issue of "harmonizing" the two regulatory approaches, particularly to reduce any
non-tariff trade barriers that otherwise might occur.
   This paper discusses and compares the U.S. and EEC requirements for new chemicals, including efforts to achieve harmony
between the two notification programs.

REGULATION OF NEW SUBSTANCES UNDER TSCA AND THE SIXTH AMENDMENT

This part compares basic provisions of the U.S. premanufacture and the EEC premarket notification programs.(3)
   Section 5 of TSCA,(4) "Manufacturing and processing notices,"
establishes the U.S. premanufacture notification program. Sections 3, 8(a)&(b), 15-17, and 19 also are important.
   Articles 5-8 and Annexes VII and VIII in the Sixth Amend-
ment(5) contain the key provisions of the EEC's premarket notification program. Articles 2, 9-13, and 20-23 also are particularly
relevant.

Basic Framework and Scope

Persons and Activities Covered. TSCA § 5 creates a premanufacture notification program, whereas the Sixth Amendment requires
the submittal of premarket notifications. Thus, U.S. PMN's must
be submitted no later than 90 days prior to the completion of
R&D activities, unless EPA grants permission to produce limited
amounts for test marketing purposes. In contrast, companies in

the EEC may manufacture, process, or otherwise produce new substances before submitting PMN's, provided they file the PMN's
prior to actually marketing the substances in the Community.
    However, the two laws are similar in that they both apply to
imported substances. In addition, they apply only to the production and marketing of new substances for commercial purposes.(6)
    Under TSCA, the first company that intends to manufacture a
new substance in the U.S. must submit a PMN for that substance.
However, once EPA completes its 90-day review of the PMN, and
after that company begins to manufacture the substance for non-R&D
purposes, the substance becomes an "existing" chemical in this
country. This means that, in the future, neither the PMN submitter nor any other company will be required to submit any further
PMN's for that substance, irrespective of any significant changes
in production volume or use, and whether any new toxicity data are
developed at a later date.(7)
    The Sixth Amendment's premarket notification requirements
differ markedly from TSCA in three important respects. First, new
substances in the EEC always will be considered "new" under the
Sixth Amendment, because the notification requirements are person-
specific -- i.e. when one company submits a PMN for a particular
substance, this does not relieve any other company from the
requirement to submit its own PMN before that second company may
place the same substance on the EEC market.(8) Second, the
Directive includes a scheme for regular follow-up reporting on
the commercial development of new substances, with progressively
more extensive (and expensive) testing requirements. Finally, the
Sixth Amendment creates a one-time notification for each company
(for each new substance), throughout the EEC. Thus, once a Member
State has completed its review of a company's PMN (without taking
any action to require further testing or to impose limitations
upon production or use), that company is not required to provide
a PMN to any other EEC country in which it subsequently markets
the substance.(9)

    Chemicals covered. Both laws exclude certain categories of
chemicals from their PMN requirements, primarily those substances
that are covered by other existing health and environmental laws.
These include pesticides, drugs and medicinal products, and radio-
active materials.(10)
    In addition to these general exclusions, neither TSCA nor the
Sixth Amendment require PMN's for new mixtures or preparations,
which generally are defined as combinations of substances that
do not result from chemical reactions. However, both laws effectively require PMN's for the marketing of new mixtures that are
new (in part) because they contain new substances.(11)
    Finally, both laws provided general exemptions for (commer-
cial) R&D substances. To qualify, companies must comply with
certain statutory limitations concerning production volume, use,
and (in the case of the EEC) numbers of customers.(12)
    TSCA itself does not explicitly exempt any other categories
of chemicals. However, in its rules and policy guidance to

date, EPA has excluded a majority of site-limited intermediates.
Further, the Agency presently is developing rules under §5(h)(4)
to exclude most polymers, the remaining site-limited intermediates, and a broad category(ies) of low-volume substances. EPA
also has indicated its willingness to entertain petitions for
similar exemptions (either chemical-specific or for general
categories) from the PMN requirements.

Because the Sixth Amendment itself exempts most of the chemicals that are subject to EPA's current rulemaking, in general the
Commission does not need to commence any exemption activities
analogous to EPA's efforts. Thus, the EEC's premarket program
covers only those new polymers that contain 2% or more of a
monomer(s). (Any new monomer is subject to the notification
requirements.) Further, because PMN's must be submitted only for
new substances that are "placed on the [Community] market," the
EEC's PMN requirements generally do not apply to the manufacture
and use of intermediates (or of any other new substances, for that
matter) by one company at one site.(13)

Likewise, the Sixth Amendment differs substantially from TSCA
in its exemption for "substances placed on the market in quantities of less than one tonne per year per manufacturer." Art.
8(1). This exemption is person-specific, but contains no time
limitations -- i.e. a manufacturer may qualify for the exemption
and thus avoid the PMN requirements indefinitely, so long as the
company does not market more than one tonne annually.(14). This
low volume exemption is contingent upon the manufacturer providing
a limited notice to each Member State in which the substance is
marketed, and complying "with any conditions imposed by those
authorities" in the various Member States.

Neither TSCA nor the Sixth Amendment exclude or exempt substances from the PMN requirements just because they are produced
by small companies. However, many of the exemptions presently
being considered by EPA, and several of those contained in the
Sixth Amendment, mitigate the burdens that the PMN programs otherwise impose upon small companies. This is particularly true for
exemptions based upon production or marketing volume.

Preemption. Unlike most other U.S. environmental laws, TSCA
is administered and enforced exclusively by EPA. None of its
rulemaking or compliance activities are delegated to the States,
and TSCA rules generally preempt any comparable state (or local)
laws and regulations.(15)

In contrast, as a Directive by the EEC Council to the Member
States, the Sixth Amendment is not self-implementing, and it is
not directly enforceable against individual companies. Rather,
each EEC country must implement its own premarket notification
laws, regulations, and administrative provisions. These must be
consistent with the overall framework of the Sixth Amendment, and
they must not create the types of conflicts or intra-EEC barriers
that the Directive was intended to prevent in the first place.
Nonetheless, they may be different from one another, to reflect
local policies and approaches to specific regulatory matters.

Initial premarket implementation activities have demonstrated
that this construct of one overriding directive implemented
through ten national laws and regulations creates many technical,
scientific, and legal problems for companies intending to market
new substances in the Community. Further, in contrast to most
other Council directives (including the other ones dealing with
health and environmental matters), successful implementation of
the EEC's PMN program may require the development of definitive
EEC-wide guidance on a number of critical matters.(16)

Contents of PMN's, Including Testing

Testing and Other Risk Data.(17)   TSCA § 5(d)(1)(B)&(C)
required each PMN to contain any (i.e. all) health and environmental effects data that the notice submitter has in his "possession or control," as well as a "description" of other data that
are "known to or reasonably ascertainable" by him. Thus, a company
must provide EPA all test data that it has developed or otherwise
obtained concerning its new chemical, and must inform the Agency
concerning any other similar data of which it is aware.   However,
TSCA does not require companies to perform tests or otherwise
develop specific data, as a prerequisite to the submission of
PMN's.(18)

To date, EPA has done three things to encourage and, in some
limited cases, require companies to test their new substances.(19)
First, from time to time EPA has issued reports, published
speeches, prepared Congressional testimony, and otherwise publicized its view that many PMN's lack necessary data to adequately
assess the subject chemicals' health and environmental effects.
Second, on a PMN-by-PMN basis EPA has negotiated with individual
companies to provide additional data and analyses. And third, EPA
has initiated actions under § 5(e) to require additional testing
for a very small number of new substances (and to limit or totally
prohibit production and use of these chemicals).(20)

The Sixth Amendment is considerably different from TSCA concerning its requirements for companies to develop toxicity information and other test data on their new chemicals. In general,
PMN's in the EEC must include the results of a required base set
of tests concerning physicochemical properties, acute toxicity,
screening for carcinogenicity and mutagenicity, and sub-acute
toxicity. Subsequent follow-up reports may trigger additional
testing requirements involving sub-chronic and chronic tests
for long-term health and environmental effects. The testing
provisions do not contain decision criteria or other rules for
determining specific testing programs. Further, the Directive
includes language that may enable companies to avoid testing in
certain circumstances.

Article 6(1) states that each PMN must include a "technical
dossier" which supplies "the information necessary for evaluating
the foreseeable risk, whether immediate or delayed, which the substance may entail for man and the environment." This includes the

results of the studies referred to in Annex
VII, together with a detailed and full description of the studies conducted and of the
methods used or a bibliographical reference
to them . . . .

Annex VII specifies six categories of information, four of which
are particularly relevant as a "base set": Category 1 - Identity
of the Substance; Category 3 - Physico-Chemical Properties of the
Substance; Category 4 - Toxicological Studies;(21) and Category
5 - Ecotoxicological Studies. Within each of these categories,
the Annex lists specific information requirements that must be
met.(22)

Neither Article 6(1) nor Annex VII contain any exemptions
from the need to perform the entire battery of base set tests.
However, the introductory language to the Annex contains an
"escape clause" for mitigating the testing requirements for particular substances (or classes of substances): "If it is not
technically possible or if it does not appear necessary to give
information, the reasons shall be stated." Although neither
the Commission nor individual countries have issued any guidance
concerning how this particular language will be applied, it is
most likely that countries will require any justifications (not
to test) to be based upon technical and scientific rationales,
rather than economic and other commercial considerations.(23)

Both the Sixth Amendment and several national premarket laws
and regulations authorize individual Member States to (1) contest
companies' claims that certain data (in the Annex VII base set)
are not necessary or are not technologically possible; (2) to
require the development of additional information and data; and
(3) to impose production or use restrictions pending the development of such additional data. Further, EEC countries may act
without having to follow many of the procedures that TSCA imposes
upon EPA, and the findings necessary to take these actions appear
to be less stringent than TSCA requires of the Agency.

Under Article 7(1), if the country that receives a notification concludes that further data and information are needed for
performing health and environmental assessments of the new substance, it may require the notice submitter to provide those data.
This may involve completion of the Annex VII "base set" (for PMN's
that invoke the "escape clause"), and/or performance of further
tests specified in Annex VIII, in addition to those contained in
Annex VII.(24) Annex VIII specifies a series of sub-chronic and
chronic tests, as well as other extensive (and expensive) data
requirements that may be required as a part of followup notifications once a chemical enters commercial production and its production volume increases substantially.

The Sixth Amendment does not elaborate upon either the substantive or procedural aspects of this authority. Rather, it
simply states that if a country can justify the need for the data
(in terms of risk assessment), it may ask for the additional

information. Further, the Directive is silent upon the relationship between this request for additional information (and any
response by the notifier to the request), and the running of the
45-day premarket review period.

Taken in conjunction with the "escape clause" in Annex VII
(which also appears in Annex VIII), this general authority to
request further information and data appears to give the Member
States considerable discretion and flexibility in their review
of PMN's and their negotiations with individual companies. Over
time, this could lead to significant differences between the PMN
requirements which companies actually face in the various EEC
Member States.

Information Concerning Production, Use, and Commercial Development. TSCA § 5(d)(1)(A) requires each PMN to include information concerning the new substance's (proposed) categories of use,
production volumes (by category of use), byproducts from production and use, estimated exposure to workers, and disposal methods.
As with toxicity data, this production and use information must
be provided only to the extent that it is "known to or reasonably
ascertainable" by the notice submitter. Because the PMN is submitted prior to manufacture for non-R&D purposes, most of this
information will be prospective in nature, and it therefore will
be expressed either in estimated ranges (where it is quantifiable
at all) or in other, more qualitative terms.(25)

Article 6(1) of the Sixth Amendment requires PMN's to contain
information and data necessary for evaluating the potential risks
of new substances to humans and the environment. This specifically includes certain exposure information listed in Annex VII,
concerning proposed uses and estimated yearly production volumes
(in ranges, and broken down by use categories). Further, Article
6(1) requires submission of "a declaration concerning the unfa-
vourable effects of the substance in terms of the various uses
envisaged," which appears to require statements of the risks that
may be associated with the use categories provided under Annex
VII.(26)

Information Concerning Classification and Labeling itself,
and Recommended Precautions and Emergency Measures. TSCA does
not contain any general requirements for the packaging and labeling of hazardous substances. Therefore it is not surprising that
there are no such requirements or similar terms for new chemicals.(27) Of course, many companies will apply the same general
hazard warnings and packaging standards to their new substances
that they use for their existing chemicals and products. This
includes the use of labels and data sheets recommended by the
American National Standard Institute (ANSI).

On the other hand, for more than a decade the EEC has
required certain dangerous substances to be packaged and labeled
according to requirements contained in this same Directive.(28)
Article 5(1) thus requires companies to package and label their

new substances in accordance with the Directive's general requirements for existing substances. Further, PMN's must include information concerning "the proposed classification and labeling of the
substance in accordance with this Directive." Art. 6(1). Member
States may use this information to review labels and require any
changes in them.
      Both Article 6(1) and Annex VII require notifiers to provide
information concerning recommended methods and precautions for
safe handling, storage, use, and transport of their substances.
Annex VII also specifies that notifications must include emergency
measures in the case of "accidental spillage" or "injury to persons (e.g. poisoning)."

Regulation of New Chemicals

   TSCA "Unreasonable Risk" Regulations, and Other Restrictions.
During the three years that companies have been submitting PMN's
to EPA, the Agency has developed a number of means for informally
regulating the production and use of certain new chemicals. In
addition, § 5(f) authorizes EPA to initiate more formal regulatory
actions, primarily involving lawsuits.
      Because notices for many new substances do not contain suffi-
cient information and data for evaluating their toxicities (especially re chronic effects) and probable exposure patterns, EPA's
primary focus in reviewing PMN's has been to determine whether it
will request further testing. In some cases, the exercise (or
threat of exercise) of the Agency's authority has proven sufficient to persuade a company either to hold up production of the
substance voluntarily (while further data are developed), or to
cease altogether its plans for bringing the chemical to market.
      Production and use of new substances can be limited in a
number of other ways as well. First, the only duty to submit a
PMN may impose costs and highlight unanswered scientific and technical questions sufficient to persuade a company that, on balance,
it should drop the substance. In addition, EPA officials often
will discuss, negotiate, and otherwise "jawbone" notifiers to
ensure that adequate production and use limitations are imposed
to protect against any significant risks that might occur. These
informal requests may be given added weight by the Agency's decision to extend the usual 90-day notice review period for up to an
additional 90 days (under § 5(c)). In fact, in a small number
of cases the submitters have agreed to suspend the notice review
period indefinitely, while further discussions and data development activities are pursued.
      In addition to these informal means for regulating, § 5(f)
authorizes EPA to seek restrictions upon the production, distribution, use, and disposal of new substances that "present or will
present an unreasonable risk" of injury to health or the environment. To ban a new chemical outright, the Agency must obtain an
injunction from a U.S. district court. Any other restrictions

must be imposed via an expedited rulemaking that is similar to the
general rulemaking actions under TSCA § 6 for existing substances.
For several reasons, EPA to date has neither initiated,
nor even seriously considered taking, any actions under § 5(f).
First, the required finding of an "unreasonable risk" to health or
the environment is very difficult to make for most new substances,
simply because of the general lack of information and data that
are needed for assessment and evaluation purposes. Second,
because both court actions and rulemakings require a significant
expenditure of Agency time, personnel, and other resources, the
procedural aspects of § 5(f) impose considerable restraints upon
the exercise of this authority. Finally, EPA has expressed its
concern about the propriety of regulating individual chemicals
under § 5(f), when those chemicals probably belong to broader
classes of substances ("me-too") chemicals) that not only present
many of the same risks to health or the environment, but do so on
a much broader scale.(29)
    In sum, EPA never has exercised its § 5(f) authority and it
probably will do so only rarely, if at all, in the future.

Regulation Under the Sixth Amendment. Consistent with the
requirement that new substances must comply with the Sixth Amendment's general provisions for the packaging and labeling of dangerous substances, the country which receives a PMN may amend the
notifier's labeling scheme to ensure such compliance. In addition, if the country requests further information and testing, it
also may "take appropriate measures relating to safe use" to the
extent that these are "necessary for the evaluation of the hazard
[i.e. risk]" that the substance may cause, and pending Community
action. Art. 7(1). But beyond these two general authorities,
the Directive does not contain a provision comparable to TSCA
§ 5(f).(30)
    Rather, Article 23 authorizes individual countries to impose
prohibitions or other special conditions upon the marketing of new
substances, pending Commission or Council action. However, this
authority is restricted by three conditions. First, the country
must have

            detailed evidence that [the] substance,
            although satisfying the requirements of [the
            Sixth Amendment] constitutes a hazard [i.e.
            risk] for man or the environment . . . .
Second, the prohibition or limitation applies only to marketing
within the country's own territory, so that a company would be
free to market its substance in other EEC countries unless they
also enact similar restrictions. Finally, any actions under Article 23, whether taken by the country of notification or another
Member State, are "provisional" -- i.e. they may be imposed only
pending Commission or Council action on the substance.

In this regard, although the country that receives the original PMN has the primary responsibility for reviewing the new substance, other Member States may participate in that review and take regulatory actions. The country that receives a PMN must provide the Commission a copy (or summary) of the PMN, and the Commission then must forward the PMN information to the other Member States. Upon receipt of this information, any other country may act under Article 23 to prohibit or limit sale of that substance in its own territory. However, this second country's action must be based upon a finding with "detailed evidence" that the substance constitutes a risk to humans or the environment in that country.(31)

Followup Reporting and Testing Requirements

TSCA and the Sixth Amendment are fundamentally different in their approaches to "followup" reporting and testing of substances for which PMN's have been submitted. TSCA § 5 does not establish such a program, and although EPA has consistently indicated its intention to require certain types of followup reporting, the Agency has yet to implement any such scheme. Conversely, followup reporting is an integral part of the Sixth Amendment, and the Member States may apply their testing and regulatory authorities (over PMN's) to these subsequent notifications as well.

Followup Reporting Under TSCA. Since 1978, EPA repeatedly has stated its intention to require followup reporting for certain PMN substances once they complete the 90-day notice review period and enter commercial production. The Agency has offered several reasons for implementing a followup program.

First, in cases when EPA lacks sufficient information and data to evaluate the potential toxicities and exposures of new substances, the Agency may face an either/or decision (either initiate a § 5(e) action or take no action), where a middle ground of followup reporting would be more desirable. Second, because TSCA PMN's are not person-specific, any "voluntary" agreements reached with the original notifiers (for example, concerning the development of additional data or the imposition of certain production or use restrictions), are not binding upon any other companies that subsequently produce the substance. Followup reporting or notification requirements applicable to all such parties (or to the major ones) might help solve this problem. Third, because TSCA establishes a premanufacture notification scheme, and not a registration or approval program, EPA often faces severe time constraints upon its ability to fully assess notices. Some type of future reporting could provide a release valve of sorts from these limitations upon the Agency's ability to perform an adequate review of those substances.

Finally, by tying certain types of followup reporting to the development of additional information and test data, EPA might be able to more realistically take account of the economic constraints facing manufacturers when they submit their PMN's (particularly small companies or producers of specialty chemicals). In fact, some U.S. companies and trade associations have suggested that EPA could adopt a type of "pay-as-you-go" approach to the PMN program -- i.e. EPA would accept minimal levels of information and data in the original PMN's, but subsequently would obtain additional information for certain "problem" chemicals once they achieve increased production volumes (and thus can justify the costs of the testing). However, EPA and industry never have concurred on some of the most basic provisions of such a program -- including the criteria for identifying chemicals for tracking in this manner, as well as the appropriate means for ensuring followup testing and reporting (e.g., voluntary agreements, rules).

EPA has cited two statutory authorities that it might use for followup. First, § 5(a)(2) authorizes the Agency to issue significant new use rules (SNUR's), which would require § 5 notifications for substances (designated in the SNUR's) when certain exposure-related "triggers" or criteria are met (e.g., significant changes in production volume). The basic concept would be that a substantial change in exposure(s) (resulting from a new use) may lead to a significant new risk(s), and it thus merits EPA's review under § 5. SNUR notifications would be subject to the same basic data requirements and review authorities described above for PMN's.

Second, EPA could issue rules under § 8(a) to require periodic reports concerning the commercial development of certain new substances once they enter production. Unlike SNUR's, § 8(a) requirements would not prevent companies from continuing their production and marketing activities. Rather, EPA would review information contained in the § 8(a) reports, and then could pursue control actions under its other TSCA authorities for regulating existing chemicals (i.e. § 4 test rules, or § 6(a) "unreasonable risk" regulations).

To date, EPA has proposed only one SNUR for a PMN substance, and the Agency has not issued any § 8(a) rules. For the most part, EPA has indicated its intention to use both of these authorities sparingly and, in particular, to build any major followup reporting efforts around § 8(a), with only very limited use of SNUR's. However, because both of these authorities require rulemaking, and for a variety of other reasons, it appears highly improbable that EPA will use either of them to any significant degree during the next several years.

<u>Followup Reporting and Testing Under the Sixth Amendment.</u>
Followup reporting and testing are integral parts of the Sixth Amendment's premarket notification scheme. The Directive itself

contains criteria for such reporting, and it specifies a variety
of tests that individual Member States may impose once they
receive the subsequent notices.

Under Article 6(4), any company that previously submitted a
PMN for a substance must inform the appropriate government concerning any of the following: significant changes in annual or
total quantities marketed; new toxicity data; new uses for which
the substance is marketed; or changes in chemical properties
resulting from a modification of the substance. Although reports
under Article 6(4) do not automatically require the development of
more toxicity information and test data, under Annex VIII, Level
1, the Member State that receives the notification may require
additional testing if production volume exceeds ten tonnes per
year, or a total of fifty tonnes, and if other relevant factors
(e.g., existing test data, uses) justify the need for more toxicity data. Several studies may be required under this authority:
fertility, teratology, sub-chronic and/or chronic (from 90 days
to two years), mutagenesis, ecotoxicological. Further, although
Annex VIII contains the same "escape clause" that is found in
Annex VII, as with base set testing the country probably can over-
rule the company's objections and require the advanced testing.

Annex VIII also contains "Level 2" notification testing
requirements, triggered if a PMN notifier's production of a new
substance subsequently reaches 1,000 tonnes per year or a total of
5,000 tonnes. The company must provide a followup notice and the
country then must draw up a test program for the notifier. This
may include tests for chronic toxicity, carcinogenicity, fertility
(e.g., three-generation study), teratology (non-rodent species),
acute and sub-acute toxicity (on second species), additional toxi-
cokinetic characteristics, and various ecotoxicological factors
(including accumulation, degradation, and mobility).

The same basic packaging and labeling requirements described
above for PMN's, and the authority for provisional regulation
of new hazardous substances, also apply to the Member States'
review of followup notifications. Thus, if subsequent notices
and additional test data warrant changes in packaging and labeling provisions, or justify the imposition of production or use
restrictions, the Member States may take action to impose such
requirements.

ACTIVITIES OF THE ORGANIZATION FOR ECONOMIC COOPERATION AND DEVELOPMENT

During the last five years, the Organization for Economic
Cooperation and Development (OECD) has directed several programs
and activities concerning the regulation of new chemicals. Most
of these have been organized under the general goal of achieving
international "harmonization" in the control of chemical substances. This part summarizes the OECD work.

The Institutional Framework

The OECD was established by Convention in 1960,(32) and grew
out of the Marshall Aid Plan's Organisation for European Economic
Co-operation. It has a membership of 24 countries, including the
U.S., all ten members of the EEC, Scandinavian countries, Canada,
Japan and Australia. In addition, the EEC "takes part" in OECD
activities.
The OECD Council is the organization's main governing body.
It consists of ambassadors or ministers from each of the member
countries. The Council usually reaches one of two types of coop-
erative agreements with regard to the major subjects that come
before it.(33) Council Decisions are binding upon all Members,
who must implement them in accordance with appropriate national
procedures and requirements. Recommendations, on the other hand,
are not binding, but are submitted to the Members who then must
decide whether to implement them through their own national laws.
Thus, the exact status of a particular Council action is significant insofar as whether it requires, or only suggests, implementation at the national level.
The OECD performs its work through various committees and
groups, with representatives designated by the participant countries. In addition, specific technical issues often are addressed
by "expert groups," comprised of specialists from the particular
countries involved. The OECD's offices and staff, located in
Paris, provide much of the day-to-day continuity and support for
Organization activities.

The OECD Chemicals Programme

The OECD Environment Committee has general supervisory
responsibilities for environmental and health matters. This
Committee supervises the work of the Chemicals Group, which in
turn has overall responsibility for issues involving the control
of substances. The OECD Chemicals Programme has two parts.

Part I Programme. As a result of Council Recommendations
in 1974 and 1977, for almost five years now the Chemicals Group
has been responsible for a number of activities which together
comprise Part I of the Chemicals Programme. Three of these are
of particular relevance to this paper: (1) The production and
subsequent updating of OECD test guidelines (containing standard
methods for performing various health and environmental tests, but
not including either rigid protocols or decision rules/criteria
concerning the circumstances when the tests should be performed);
(2) The development of approaches to "step sequence" testing;
and (3) Work on principles and guidelines for hazard (i.e. risk)
assessment.(34)

Part II Programme. In 1978 the OECD instituted its Part II Programme, the Special Programme on the Control of Chemicals. This work is supervised by the Management Committee.(35)

To date, the Part II Programme has focused upon four major projects: (1) The development and implementation of a set of Principles of Good Laboratory Practice (GLP's); (2) Resolution of issues concerning Confidentiality of Data; (3) Development of a Glossary of Key Terms; and (4) Development of guidelines and other procedures for the exchange of information (e.g., re test data, the export of hazardous chemicals, and the labelling of hazardous chemicals).

Work To Date. From 1977-80, the most intensive and productive OECD activities focused upon Mutual Acceptance of Data (MAD) and the development of test guidelines and GLP's. Efforts also were devoted to the Step Sequence Group and, in particular, that body's efforts to develop a Minimum Pre-Marketing Set of Data (MPD). Technical and scientific work also progressed on the various hazard assessment issues; and expert groups worked on recommendations concerning confidential data, definitions of key terms, and principles of information exchange.

In May 1980, the Chemicals Group endorsed recommendations from three of its groups concerning GLP's, test guidelines, and the MPD, and endorsed the principle of Mutual Acceptance of Data. Thereafter, the Environment Committee also endorsed these recommendations.

A year later, in May 1981, the OECD Council considered these recommendations and issued a Decision establishing the following principle of Mutual Acceptance of Data (MAD):

> [D]ata generated in the testing of chemicals
> in an OECD Member country in accordance with
> OECD Test Guidelines and OECD Principles of
> Good Laboratory Practice shall be accepted in
> other Member countries for purposes of assess-
> ment and other uses relating to the protection
> of man and the environment.

In Annexes accompanying its Decision, the Council incorporated the specific GLP's endorsed by the Chemicals Group, as well as those test guidelines that had been developed to date.(36) The Council did not endorse any action concerning the MPD.

It should be noted that the Council's Decision does not commit members of the OECD to adopt these test guidelines and GLP's as enforceable requirements. Rather, the Council issued a Recommendation that "Member countries, in the testing of chemicals, apply" the guidelines and GLP's. Thus, while the Decision commits Members to accept, for purposes of their own assessments and evaluations, any data that are generated in accordance with these guidelines and principles, Members remain free to select these or any others for their own national use.

The Minimum Pre-Marketing Set of Data (MPD)

Perhaps the most controversial aspect of the OECD's work concerning new chemicals has been its efforts to develop a Minimum
Pre-Marketing Set of Data, sometimes referred to as a "base set".
This activity is the first of the Step Sequence Group's efforts
to identify principles and criteria for determining when various
tests should be performed. As distinguished from the work to
develop GLP's and test guidelines, this effort has focused upon
the application of those guidelines to different types or categories of chemicals and in various production and use situations.
Thus, it raises the fundamental issue of so-called decision rules,
criteria, and other similar guidelines for deciding under what
circumstances the specific tests should be performed.
    As noted above, in May 1980 the High Level Meeting of the
Chemicals Group endorsed the MPD, with full support from the U.S.
representatives. The data components of the MPD were largely
similar to those contained in the base set of tests found in
Annex VII to the Sixth Amendment. The MPD also would include
"provisions for flexible applications" of the MPD to particular
chemicals, using language similar to the "escape clause" found in
Annex VII. The draft Council Decision read, in pertinent part:
        [The MPD] shall be generated or obtained and
        applied for the purpose of initial assessment
        of new chemicals, and in this regard the data
        components of the [MPD] and the provisions for
        its flexible application are set forth in the
        Annex to this Decision and form integral parts
        thereof.
The Environment Committee endorsed the MPD and recommended it to
the Council for action at its May 1981 meeting. However, the
Council, in the face of new U.S. opposition to enactment of the
MPD as written, failed to enact either a Decision or a Recommendation concerning MPD.
    This change in official U.S. policy came as a result of lob-
bying that the American chemical industry (primarily the Chemical
Manufacturers Association (CMA)) undertook with the new Administration after it assumed office in January 1981. The major concern that CMA stated was that the proposed Council Decision on
MPD, as written, might legally bind the U.S. to amend TSCA and
require the MPD as part of PMN's in this country. CMA long had
opposed any concept of "base set" testing for all new substances
under TSCA, and viewed the MPD, particularly if incorporated in
a Council Decision, as being directly contrary to that position.
As a result, industry prevailed upon the (new) U.S. officials to
oppose enactment at this time of any Council measure dealing with
MPD; the U.S. representatives at the May 1981 Council meeting
opposed the draft Decision; and the Council did not act upon it.

Subsequent to the Council meeting, discussions have continued, both within this country and informally at a number of OECD meetings, to attempt to resolve the present U.S. opposition to the earlier draft Decision. Attention recently has focused upon a draft "interpretive statement" to be added to the original draft MPD Decision, which would make it clear that enactment of the Decision in no way would bind the U.S. to amend TSCA or otherwise incorporate MPD into the U.S. PMN program. However, no consensus has been reached to date.

HARMONIZATION AND OTHER FUTURE DEVELOPMENTS

TSCA and the Sixth Amendment are quite different in many important respects, and it would require some fundamental changes to standardize them or even make them consistent. However, because they deal in part with the same general subject matter -- industry notification and government review of new chemicals -- and because they may lead to trade barriers and the inefficient use of scarce technical and scientific resources, it is useful to consider how they might be brought closer in line with one another. "Harmonization" is a term commonly applied to such efforts.

The Concept of Harmonization

Webster defines "harmony" to mean "correspondence, accord . . . [as in] 'lives in harmony with her neighbors.'" Thus, to "harmonize" is "to bring into consonance or accord." The term "accord" is defined as "balanced interrelationship: harmony."(37)
None of those persons active in international discussions concerning the regulation of new chemicals has argued that "harmonization" will result in the development of any standardized, world-wide scheme for the submission and review of PMN's. Rather, most have viewed harmonization as being goal-oriented -- in the words of one OECD official,

> Harmonization is something more than coordination -- which doesn't convey the idea of shared objectives -- but something less than standardisation -- since there is generally a variety of acceptable ways to attain agreed goals.(38)

Thus, given diverse legal and economic frameworks, efforts to achieve harmony on major policies and procedures depend upon the various parties reaching accord on their basic goals.
With regard to the testing and regulation of new chemicals, it is not at all clear that, with the exception of the most fundamental goal of protecting man and the environment from certain chemical risks, the major parties involved in these issues have reached an accord on fundamentals. In fact, experience to date indicates that real "harmonization" often neither reflects nor

constitutes a goal of multilateral discussions, but is cited in support of other desired outcomes once they have been achieved through such discussions.

Of course, ongoing dialogue between representatives of the various governments and industries involved may lead to a common understanding of the different regulatory frameworks and of the various parties' interests in them. But beyond this most basic outcome, there appear to be at least four levels of concurrence at which some sort of "harmony" could conceivably be achieved concerning notification and review of new substances: (1) The development of common methodologies and guidelines (e.g., concerning toxicity tests, economic analyses), and agreements to accept the data derived from these methods; (2) Agreements to employ these methods as the recognized ("approved"?) approaches to developing the particular data and analyses; (3) Concurrence on joint criteria and "decision rules" for developing data and information (addressing the circumstances under which the methods and guidelines should or must be employed); and (4) Joint or common approaches to the actual review and regulation of new substances. Using these levels as a reference point, the following discussion evaluates developments to date and what may lie ahead.

Harmonizing TSCA and the Sixth Amendment

Conflicts Between the Two PMN Programs. Inconsistent and conflicting PMN requirements arise in each of the four areas described in the first part of this paper.

First, to the extent that TSCA and the Sixth Amendment do not have the same general scope and coverage, this necessarily will mean that companies in some cases will face regulation of their (new) chemicals and commercial activities in one country(ies), but not in another, and vice versa. Almost by definition, this will create certain artificial competitive advantages and disadvantages for companies, depending upon at which end they lie in the trade of those particular chemicals.

Similarly, different testing and other data requirements may create non-tariff barriers to trade. If countries reach agreement concerning the use of common test methods and guidelines, and if they further harmonize concerning both the use (e.g., the OECD's MAD) and sharing of information and data, some of these barriers will be reduced. But to the extent that the substantive testing criteria are at variance, either as written or as applied by national regulatory officials (e.g., re the "escape clause" in the Sixth Amendment's base set of tests), very real barriers will remain.

The inconsistent or conflicting regulation of new chemicals probably represents the most obvious manner in which the PMN laws can create barriers to trade. Given the fundamental differences between the U.S. and European PMN programs, as well as the inherent self-interests involved because the EEC and the U.S. are

competitors in the world-wide chemicals trade, it is highly
unlikely that any significant agreements ever will be reached
concerning the joint regulation of new substances. At the very
most, and only at some point in the distant future, it might be
possible to reach some basic accord concerning the factors to
be considered in making regulatory judgments (i.e. health and
environmental risks, the efficiency of control options, economic
impacts of the regulations). Perhaps the best evidence of this
is the fact that even within the EEC itself, the regulation of
chemicals -- albeit provisional -- is left largely to the discretion of individual Member States.

Finally, the different approaches to followup notification
and testing impose additional requirements upon substances in
Europe that do not apply to the same chemicals when they are
marketed in the U.S. In the short term, at least, this discrep-
ancy may well work to the advantage of companies doing business
in this country (whether U.S.- or foreign-based). But, over time
this also could lead European companies to seek changes ("harmonization"?) which either make the EEC requirements less stringent or
incorporate an EEC-type of followup scheme into TSCA.(39)

Developments to Date. It often has been stated that the
basic policy objective of efforts to harmonize the U.S. and
European laws is the achievement of consistent and effective
protection of health and the environment. However, economic
considerations -- in particular, the avoidance (or minimization)
of non-tariff trade barriers -- constitute the principal force
behind virtually all of these multilateral efforts. The trade
in chemicals and chemical products constitutes a significant
part of the overall trade between Western industrialized nations.
Specifically, the U.S. enjoys a favorable balance in its chemicals
trade, and this is particularly significant given the current
recession. Thus, any unnecessary barriers to this trade may
impose substantial burdens upon certain segments of the American
chemical industry, and may constitute violations of the international General Agreement on Tariffs and Trade (GATT).

To date the U.S. and the EEC have implemented their respective PMN programs largely independent of one another. Because
they have needed to draft basic rules and policy statements,
implement inventory reporting requirements, and staff up and
prepare for the receipt of PMN's, it is hardly suprising that
the respective governing officials have focused their time and
resources upon "getting their own acts together," with only
secondary attention devoted to reaching accord on many of the
inconsistencies and conflicts described in this paper. Further,
to participate in any meaningful discussions and negotiations
with their counterparts abroad, representatives from the several
countries and organizations must have developed their own
(initial) approaches to the many complex issues that they face.

Following passage of TSCA in 1976, and during EPA's initial
implementation of the premanufacture program in 1977-78, the U.S.
recognized the need to participate in multilateral discussions
concerning coordination and "harmonization" of its regulatory
activities. For a variety of reasons, U.S. officials selected the
OECD as the primary forum for such discussions (and negotiations).
Thus, except on a purely informational basis, the U.S. chose not
to discuss or negotiate directly with the EEC concerning implementation of TSCA and the Sixth Amendment. Further, the lead U.S.
representatives to the OECD chemicals work (including the U.S.
chairs of the Chemicals Group, the Management Committee, and of
various expert groups) were drawn from EPA and other federal
health and environmental protection agencies, and they did not
consider commercial and economic issues to be their primary concern in these efforts.

Of course, the work that has gone on to date within the OECD
demonstrates the usefulness of focusing upon relatively noncontroversial and apolitical issues upon which agreement can be reached
at an early stage. The recent OECD Decision concerning MAD, test
guidelines and GLP's represents a very constructive outcome of
these harmonization efforts, and it bodes well for the success
of similar "Level 1" efforts(40) -- including agreements concerning the glossary of key terms and the mechanisms for information
exchange. But falling short, as they do, of requiring consistency
in matters of scientific, legal, or regulatory judgment (e.g.,
whether to perform certain tests, and the use to which resulting
information and data should be put), these early successes should
not be taken as an indication that similar consensus can be
reached concerning issues requiring a higher level of understanding and agreement. The debate over adoption of the MPD is the
first, and probably not the last, evidence of this.

Prospects for the Future. During the next few years, a
number of factors -- for the most part, missing until now -- may
provide a more concrete basis for productive dialogue on major PMN
issues. These include the fact that the EEC will begin to develop
some real experience with its own program, which should enable the
Community to better understand its own system and identify those
parts in which changes are both necessary and feasible. This is
particularly true with regard to such major issues as testing and
followup notifications.

In addition, as a second generation of government officials
and legislators assume responsibility for implementing (and,
as necessary, amending) these laws, they may be more flexible
concerning major policy issues than the persons who have been
responsible for establishing the initial strategies for the PMN
programs. With the change in Administration and turnover of EPA
officials in the U.S., such modifications already are evident.
Presumably, the same scenario will develop in the EEC. Although
the political and philosophical views of the new persons obviously

will influence their approaches to the new chemicals program, the mere fact that different persons are involved should, in and of itself, set the stage for further discussions on major harmonization issues.

However, the basic differences between the U.S. and EEC laws never will be reconciled in one common approach to the regulation of new chemicals, and it is highly improbable that major multilateral agreements will be reached on key regulatory issues. Thus future efforts, both at the OECD level and involving direct U.S.-EEC discussions, will be most profitable in fostering mutual understanding about how the two systems actually operate, and then in achieving Level 1 agreements similar to those reached to date. At most, it may be possible to reach consensus on certain Level 2 issues concerning the standard analytical methods and guidelines to be employed if national laws otherwise require performance of the underlying tests and analyses. Beyond this, the prospects for achieving any high levels of harmony are, at best, pretty slim.

Rather, based upon their cumulative experiences with PMN's, and in response to domestic (including intra-EEC) economic and political factors, during the next few years the U.S. and EEC each will fine tune its own laws through a combination of regulations, policy statements, administrative decisions, and judicial actions. These adjustments might result in greater consistency between TSCA and the Sixth Amendment, but direct "harmonization" will not be the driving force behind most of these changes. Thus, both industry and government should anticipate only limited (if any) success in efforts to eliminate the major differences in the two laws' treatment of significant PMN issues.

The following discussion predicts some of the major developments concerning the U.S. and EEC PMN programs. Admittedly, it is quite speculative and subject to considerable change based upon a variety of factors. Nonetheless, it should give some idea of what we may expect during the next decade of PMN notification and review.

1982-1983. Through the end of 1983, it is unlikely that there will be any major new developments resulting in the harmonization of TSCA and the Sixth Amendment -- either from multilateral agreements as such, or from parallel developments in the implementation of the two laws. This stems primarily from the EEC's focus upon intial implementation of its own requirements, and from recent changes in the U.S.'s approach to OECD and other multilateral discussions.

There will be three major developments in the EEC: (1) reporting and compilation of the EINECS(41) inventory; (2) submission of the first wave of PMN's to the various Member States, and their subsequent review and response to those PMN's; and (3) initial work at the Commission level (in consultation with the Member States) to identify and then address major issues that

require EEC-wide discussion and resolution (including the development of some types of guidance). As a result of these three
efforts, both the Commission and the individual States will begin
to develop a track record on exactly how they will approach the
PMN program, as a basis for the development of more general policies and procedures.

The major work under TSCA will be to conclude the current,
highly-publicized efforts to develop broad exemptions from the PMN
program for polymers, site-limited intermediates, and certain low
volume substances. EPA probably will publish general PMN rules
and notification forms which essentially restate the applicable
statutory terms, and which may reflect the Agency's conclusions
concerning the minimal data that it would encourage submitters to
provide. Also, the Agency will take few (if any) actions under
§ 5(e), and will initiate only a skeletal followup reporting
program.

Finally, there will be a general weakening of the OECD's
role as the major forum for harmonizing the two PMN systems.
The present efforts concerning information exchange will continue,
as will others dealing with noncontroversial technical and scientific matters. But the OECD probably will have little success
in addressing and resolving any key regulatory or policy matters,
including the Council's consideration at MPD.

1984-1986. TSCA and the Sixth Amendment will continue to
develop largely in parallel, but a type of "harmony" may emerge
indirectly as a result of changes that the U.S. and EEC make
concerning the basic scope and coverage of their laws. Thus,
the U.S. will have completed its initial § 5(h)(4) exemption rulemakings, and the EEC will have better defined which chemicals and
activities are subject to its PMN requirements. Also, both sides
will continue to develop better methods and data bases concerning
fundamental scientific and economic issues, which in turn will
enable them to more profitably compare and discuss areas of inconsistent or conflicting regulation.

Specifically, under the Sixth Amendment companies will have
begun to refine their procedures for submitting PMN's and then
negotiating with individual Member States concerning data needs
and possible use restrictions. This, in turn, will lead some
companies to a type of "forum shopping" -- i.e. they will select
those countries in which, for a variety of reasons, it is most
opportune to submit their PMN's, and they will avoid those in
which it is not. Also, the Commission will have issued (or otherwise expressed its tacit approval of) guidance on some fundamental
technical issues, particularly concerning the contents of PMN's.
Several Member States will have taken some regulatory actions on
new chemicals, thus raising basic legal and regulatory issues concerning the extent to which the EEC can and will permit individual

Members to regulate substances (within their own territories) when
other countries (and the EEC itself) do not concur with such regulations.

In the U.S., Congress for the first time will give serious
consideration to making major amendments in TSCA, primarily to cut
back on some of its more egregious provisions and to otherwise
focus upon particular types of health and environmental problems.
EPA will continue to issue exemptions for specific categories of
substances, and will develop some general followup requirements
for a few, well-defined categories -- i.e. those for which PMN's
consistently demonstrate possible health and environmental risks,
once commercialization begins. Also, the Agency will move to
publish a more explicit set of criteria for selecting tests to
assess the risks presented by certain categories of non-exempt
chemicals, incorporating concepts of "flexibility" analogous to
the general "escape clause" found in the Sixth Amendment. These
criteria and other related guidance will reflect a consideration
of the economic constraints involved in the testing of new substances.

The OECD will continue its technical work, but its focus will
shift primarily to existing chemicals. In addition, the Organization will increasingly focus its attention upon waste management
issues, again following the approach taken for new chemicals --
emphasizing the need to reach consensus on fundamental technical
and scientific matters, as well as on the exchange of information
and data.

1987-1991. Based upon discussions conducted directly between
the U.S. and the EEC (as well as bilateral talks involving Canada
and, possibly, Japan), some "harmonization" may be reached concerning "base set" testing of new substances and fundamental
criteria for evaluating PMN's. These will not, however, be
reflected in any major treaties or conventions, or through discussions within the OECD. Rather, TSCA and the Sixth Amendment
will continue to develop in parallel, and the U.S. and the EEC
may reach accord on some very basic regulatory matters.

Within the EEC, there will be moves to regulate a few categories of (new) "problem" chemicals, primarily those that several
Members repeatedly have identified as being of concern for marketing on a Community-wide basis. Also, the Commission will recommend (to the Council) certain amendments, primarily to make
changes in the Directive that already have been accepted in
practice by Member States.

Under TSCA, EPA also will turn its attention to requiring
additional up-front testing and other data development for certain
categories of new substances that consistently lack data (in the
PMN's), and which are known to be inherently toxic. The primary
vehicle for these additional requirements will be the § 5(b)(4)
"risk list," although some § 4 test rules also may be used.
By this time, too, the Agency will have refined its means for

imposing production or use restrictions absent formal rulemakings
or court actions, and the Congress may consider amending TSCA
to incorporate some of these extralegal approaches into the Act.

FOOTNOTES

1. Paper presented to the Symposium on TSCA Impacts on Society
   and the Chemical Industry, American Chemical Society 183rd
   National Meeting, Las Vegas, Nevada, April 1, 1982 -- Blake
   A. Biles, Jones, Day, Reavis & Pogue.
2. The U.S. premanufacture notification (PMN) requirements have
   been in effect for approximately three years, and more than
   1250 PMN's have been submitted to EPA. The EEC's premarket
   notification requirements took effect in September 1981,
   so that to date only a handful of PMN's have been filed in
   Europe.
3. This discussion comparing the two PMN programs is substantially shortened from the paper presented at the Las Vegas
   ACS program. Copies of the original, more detailed text are
   available from the author.
4. Toxic Substances Control Act, 15 U.S.C. 2601 et seq., Pub.
   L. No. 94-469.
5. Council Directive No. 79/831 of September 18, 1979 (O.J. No.
   L 259/10 of October 15, 1979). This Directive is referred to
   as the "Sixth Amendment" because it represents the sixth time
   that the Council has amended Council Directive No. 67/548 of
   June 27, 1967 (O.J. No. L 196/1 of August 16, 1967) concerning Classification, Packaging and Labelling of Dangerous
   Substances.
6. EPA has construed this "limitation" quite broadly, so that
   only those activities that are carried out in government or
   university (or similar non-profit) research labs, and that
   are not tied to any subsequent commercial exploitation,
   are outside the scope of the premanufacture requirements.
   The EEC can be expected to take a similar approach to inter-
   preting what is "commercial."
7. See the discussion concerning EPA's intention to impose
   followup reporting or notification requirements upon the
   producers of some PMN'ed substances.
8. Similarly, if one person is exempt from the EEC's notification requirements, this does not relieve any other person
   from the requirement to submit a PMN if that second person
   is not also entitled to the exemption.
9. However, multiple PMN's may be required for substances manufactured in more than one EEC country, due to the incorpora-
   tion of several legal entities in the various Member States.

10. Other exclusions under TSCA include tobacco and tobacco products, firearms, and other substances and products that are regulated under the Federal Food Drug and Cosmetic Act. The Sixth Amendment similarly excludes several broad categories of substances and activities: narcotics; the transport of dangerous substances by rail, road, inland waterway, sea or air; wastes that are regulated under other Council directives; substances that are imported into the EEC and then exported without being processed or used in any manner while in the Community; fertilizers that are regulated under other EEC requirements; and substances that are "already subject to similar testing and notification requirements under existing Directives."

11. However, EPA has stated that the TSCA PMN requirements do apply to the import of new substances as a part of mixtures. If the premanufacture requirements only applied to the import of new substances in bulk, companies could avoid the PMN requirements by producing new (to the U.S.) substances abroad, formulating them into mixtures abroad, and then importing the new mixtures into this country.

12. Article 8(1) of the Sixth Amendment specifies three different types of exemptions for R&D substances. The third appears to be analogous to the TSCA § 5(h)(1) exemption for test marketing activities.

13. However, new substances that are imported into the Community and then used at the site of import are subject to the EEC PMN requirements.

14. Further, because the volume restriction relates to marketing rather than manufacturing, it appears that a company is entitled to manufacture several tonnes of a chemical in one year, and then market it over a period of several years (at less than one tonne per year), without being required to submit a PMN for the substance.

15. See TSCA § 18, "Preemption."

16. Examples of these issues include the current uncertainty over the scope of the Directive's general exemption for polymers; the precise interpretation of the exclusions for pesticides and medicinal products; conflicting national approaches to the definition and protection of confidential business information; and differences of opinion (between individual Member States) concerning the applicability of the PMN requirements to substances previously marketed in certain countries but not in others.

17. It is important to note the meaning of two terms used in con-
    nection with TSCA and the Sixth Amendment, because they often
    have different interpretations under the two laws. "Hazard"
    generally is used in the Sixth Amendment (and in other EEC
    directives) to mean what U.S. scientists and regulators often
    call "risk" -- i.e. an assessment or evaluation that consid-
    ers both the effects and exposures that are associated with
    particular substances. In this country, the term "hazard"
    usually refers to the inherent toxicity or effects of a
    substance, or to the unsafe characteristics of particular
    chemicals or products. The latter meaning is the one used
    in this paper.

18. In January 1981, EPA published a Premanufacture Testing
    Policy containing general test methods and guidelines for a
    large number of human and environmental effects. 46 Fed.
    Reg. 8986, Jan. 27, 1981. Although this guidance would not
    have imposed any legal obligation upon companies to conduct
    tests for those effects, it endorsed the OECD's "Minimum
    Pre-Marketing Set of Data" (MPD), a "base set" of tests to
    be performed on most new chemicals. However, following the
    change in Administration in 1981, the U.S. withdrew its
    active support for the MPD. For a further discussion of
    this issue, see the next part of this paper.

19. EPA possesses a fourth means for obtaining additional test
    data from industry, although the Agency has never used this
    authority. Under § 4(a), EPA may impose testing require-
    ments for categories of chemicals, provided the Agency makes
    certain findings about the need for such data and about the
    chemicals' potential for exposure to human or environmental
    populations. Section 5(b)(1) provides that any PMN for a new
    substance covered by such a testing category must contain the
    requisite test data. Taken together, these provisions autho-
    rize EPA to impose testing requirements upon new chemicals in
    certain classes, as a condition for the submittal of PMN's
    for those substances.

20. Under § 5(e), following receipt and review of a PMN EPA may
    order a company to develop test data "sufficient to evaluate
    the health and environmental effects" of the new substance.
    However, if the PMN submitter objects to the order (and
    provides sufficient grounds for that objection), the order
    does not take effect and EPA must obtain an injunction from
    a U.S. district court to impose the data requirements (and
    any appropriate production or use restrictions).

21. Category 4 lists three types of studies for human health
    effects: basic acute toxicity tests, a 28-day animal study
    (referred to in other discussions as a "sub-chronic" test),
    and a series of two (or more) screening tests for mutagenic-
    ity and carcinogenicity.

22. The Directive contains these other requirements concerning toxicity information and test data: (1) Notifications must include descriptions of the studies conducted and methods used (Article 6(1), and Annex VII, Introductory Statements); (2) The tests must be performed according to the methods specified in Annex V (Article 3(1)), and must be "recognized and recommended by the competent international bodies where such recommendations exist" (Annex VII, Introductory Statements); (3) The persons who carry out the tests must comply with the principles of current good laboratory practice (Annex VII, Introductory Statements); and (4) The notifications must include the composition of samples used in testing, and the name of the persons responsible for carrying out the studies (Annex VII, Introductory Statements).

23. Under Article 6(2), if a particular substance previously was the subject of a PMN by another company, the notifier may rely upon data submitted in the prior notification(s). However to take advantage of this "me-too" provision, a company must obtain agreement (1) from the relevant government authority that reference to the earlier results is satisfactory, and (2) from the previous submitter to use that submitter's data.

24. For a discussion of the authority of other Member States to request additional testing, see footnote 31.

25. Upon receipt of a PMN, EPA has little direct recourse for requiring a company to develop and otherwise provide more production and use information. Section § 5(e) is EPA's major legal authority for obtaining additional information, and that section focuses upon health and environmental effects data, rather than exposure information.

26. The Directive does not specify any other exposure-related information (e.g., production processes, effluent and emission data, information on worker exposures), and it does not mention economic or other non-risk information. However, individual Member States may attempt to require this type of information under their general authority to require additional information and data beyond those specified in Article 6(1) and Annex VII.

27. Although TSCA § 6(a)(3) authorizes EPA to issue requirements containing hazard warnings and instructions, the Agency never has proposed any such rules. This may be due in part to the fact that OSHA has proposed certain hazard communication requirements (47 Fed. Reg. 12091, March 19, 1982) which may become effective during 1983.

28. As noted earlier, the "Sixth Amendment" enacting the premarket notification requirements constitutes the sixth time that the EEC has amended the Directive concerning the Classification, Packaging and Labelling of Dangerous Substances. This Directive was enacted on June 27, 1967 (O.J. No. L 196/1 of August 16, 1967). The key provisions concerning packaging and labeling in the Directive (as now amended) are Articles 16-18 and Annexes I-IV; also see Articles 2-4, 14-15, and 20-23.

29. This "me-too" issue also has been debated in the context of possible § 5(e) actions, concerning the merits of requiring tests for new chemicals when their broader classes of substances are not also being tested.

30. Of course, the impact even of temporary measures can be significant, particularly if the time it takes for the Community to take action, and for the notifier then to comply with that action, is long relative to the notifier's need to introduce the chemical onto the market. Also, this regulatory authority may provide considerable leverage to any data-gathering requirements imposed under Article 7(1), just as prospective TSCA § 5(e) actions can have a significant impact upon notice submitters in the United States.

31. Other Member States also may use the PMN information and data that they receive as a basis for requesting further tests or information from the PMN submitter (in addition to any such requests from the country that receives the PMN). However, the second country first must suggest to the notified country that the latter request the further testing or other information from the company. If the two countries then disagree concerning data to be requested from the company (or disagree on the need to request any additional data at all), either country may ask the Commission to request such information. Thus, the primary responsibility for requesting additional information and testing lies with the country of notification, and to have an impact, any other Member State must convince either the original country or the Commission that the additional data must be requested.

32. Convention of the Organisation for Economic Co-operation and Development, December 14, 1960.

33. The Council also may enter into agreements with Members, nonmember states, and international organizations.

34. The other major elements of the Part I Programme are an Economics Programme (studies and other analyses of the economic impacts of assessment procedures, guidelines, regulations, and laws); the Complementary Information Exchange Procedure (concerning the exchange of information about various laws and regulations); and concerted actions on specific chemicals (re PCB's and mercury, to date).

35. The Part II Programme was organized under a separate Management Committee (rather than the existing Chemicals Group) in order to obtain the necessary funding, but not to establish a separate supervising committee. The Management Committee and the Chemicals Group closely coordinate their activities and have overlapping memberships.

36. At the time of the Council Decision, approximately 50 guidelines were in final form. An equal number were in various stages of drafting, and are to be incorporated into the Annex as they are completed.

37. Webster's New Collegiate Dictionary, 8, 519 (1981 ed.).

38. "OECD Harmonization of Chemicals Control," p. 2, Speech by B. Gillespie, Administrator, OECD Chemicals Division, presented to CMA/IAG Seminar on Compliance with International Chemical Regulations, Washington, D.C., April 29, 1981.

39. In addition, the requirements and policies concerning the handling and treatment of confidential business information (CBI) are and will continue to be an important topic of "harmonization" discussions. In a number of ways, different approaches to CBI (between the EEC and the U.S.) will directly affect the CBI protections available within each country(ies). This is particularly true concerning the effects that the disclosure of proprietary information in one country will have upon industry's ability to protect CBI from disclosure in other nations.

40. See my discussion above of the four "levels" of possible harmonization.

41. EINECS: European Inventory of Existing Commercial Chemical Substances. This is the inventory of existing substances required by Article 13(1) of the Sixth Amendment.

RECEIVED December 14, 1982

Control of Existing Chemicals

ETCYL H. BLAIR and CARLOS M. BOWMAN

Dow Chemical Company, Midland, MI 48640

The goal of the Toxic Substances Control Act
(TSCA) is to provide authority to regulate chemical
substances which present an unreasonable risk of
injury to health or the environment. An important
feature of TSCA requires the administrator of the
Environmental Protection Agency (EPA) to examine
such data on existing chemicals and, when it is
insufficient, to direct industry to conduct tests.
A study of the inventory list of 55,000
commercial substances shows that less than 10% of
them account for 99.5% of production. A panel of
authorities could provide a qualitative ranking of
risk uncertainties which could narrow the list of
substances to those of most immediate concern.
Those could then be the subjects of case-by-case
consideration by appropriate experts.

The goal of the Toxic Substances Control Act is to provide
authority to regulate chemical substances and mixtures which
present an unreasonable risk of injury to health or the environment. Congress recognized that in order to make a finding of
unreasonable risk the Environmental Protection Agency administrator must have adequate data to make such an assessment. One of
the important features of TSCA is the requirement that the administrator examine the health and environmental data of existing
chemicals (Section 4), and where data is found to be insufficient
to make an assessment, industry must do the testing.

The magnitude of the number of commercial substances
(55,000) and the limited resources available to examine the
potential risks posed by this large number of substances require
that we understand the universe of chemicals we are considering
so that guidelines for priority setting can be established.

We tried to obtain an understanding of the universe of
chemicals by examining the U.S. EPA TSCA inventory information
which has been made available to the public. The data compiled

0097-6156/83/0213-0067$06.00/0

in this collection include those materials submitted early in
the reporting period and omit, of course, confidential information*. Nevertheless, it is adequate to provide a good picture
of the universe of commercial chemicals.

As would be expected, we find that a small number of
materials account for the bulk of the production volume. In this
case, those materials produced in excess of 100 million pounds
per year represent only 1.8% of the total number of substances
reported and account for 98.9% of the total pounds produced.
Lowering the limit to 10 million pounds adds 2.7% of the chemicals, so we now have 4.5% of all the substances and we increase
the total volume represented to 99.7%. Going further to one
million, we find that only 9.5% of the materials account for
99.9% of the total production reported (Table I).

Table I

VOLUME DISTRIBUTION
Entire EPA Inventory

Production Range (Lbs/Yr)	Number of Materials	%	Total Production (Million Lbs/Yr)	%	Cumulative % Production
$>10^{11}$	1	<0.1	102,000	2.5	2.5
10^{10} 10^{11}	95	0.2	3,119,000	76.5	79.0
10^{9} 10^{10}	216	0.5	656,000	16.1	95.1
10^{8} 10^{9}	436	1.1	155,000	3.8	98.9
10^{7} 10^{8}	1,065	2.7	33,800	0.8	99.7
10^{6} 10^{7}	1,983	5.0	8,140	0.2	99.9
10^{5} 10^{6}	3,798	9.7	1,720	0.04	99.98
10^{4} 10^{5}	4,689	11.9	225	0.01	99.99
$<10^{4}$	27,010	68.7	28	<0.01	100.00

To better understand these high-volume materials, we divided
all those substances reported as produced in quantities of one
million pounds per year or more into several categories such as
organics, inorganics, polymers, etc. This exercise was most
revealing. We found that those materials which can be classified
as petroleum derivatives (gasoline, kerosine, distillation cuts,

* The data used in this discussion were that made available by the
EPA from the TSCA inventory and do not have production volume
information when that information was claimed confidential.
Since the production volume was reported in ranges, in this
exercise an appropriate midpoint was used for each range --
except that 1,000 pounds was used for materials reported as
produced in quantities less than 1,000 pounds, and one billion
pounds was used for those reported as being produced in
quantities larger than one billion pounds.

Table II

VOLUME DISTRIBUTION BY TYPE OF SUBSTANCE
(Produced in Excess of 1,000,000 Lbs Annually)

Type of Substance	Count	% Count	Production Volume Millions	% Volume
Petroleum, Primary Derivatives	380	10.0	2,258,000	55.4
Inorganics	452	11.9	503,000	12.4
Metals, Refining Residues (Ferrous)	20	0.5	281,000	6.9
Alkanes	21	0.5	272,000	6.7
Organics	1,307	34.4	246,000	6.0
Polymers & Plastics	893	23.5	122,000	3.0
Other	18	0.5	95,000	2.3
Coal, Primary Derivatives	30	0.8	90,500	2.2
Natural Products & Derivatives	254	6.7	84,200	2.1
Metals, Refining Residues (Non-Ferrous)	52	1.4	59,700	1.5
Organics, Variable Composition	287	7.6	27,000	0.7
Metals	24	0.6	19,600	0.5
Minerals	29	0.8	14,300	0.3
Alloys	13	0.3	1,600	0.04
Dyes & Pigments	15	0.40	133	<0.01
Living Organisms	1	0.03	1	<0.01
Total	3,796	100.00	4,074,000	100.00

etc.) represent 10% of the total number of entries in the inventory, but account for 55% of the total production. The inorganics represented 12% of the materials and 12% of the production. Another 7% of the production is due to materials which are residues from the processing of ferrous metals. The saturated hydrocarbons (methane, ethane, hexane, etc.) were responsible for 7%.

We found that structurally well-defined organic substances, the materials we are most concerned about in testing, were the most numerous as they represented 34% of the inventory sample, but they account for only 6% of the total production. Polymers and plastics represent 24% of the number of materials and 3% of the total production (Table II).

Some institutions are giving some type of attention to the petroleum refining, metal or metallurgical substances. However, it appears that the U.S. EPA has either consciously or unconsciously leapfrogged those materials and began by focusing attention on organics and some of the inorganics. Thus, attention is being centered on some 1,307 substances which represent about 6% of the U.S. total production volume.

The organic grouping exhibits the same volume pattern of production as the inventory as a whole. There were 3.3% of the organics produced in quantities in excess of a billion pounds, and this group represents 77% of the total organic production (34% of total). The 100-million to one-billion pound range is 9.2% of the number and an additional 17.2% of the volume (Table III).

Table III

VOLUME DISTRIBUTION OF ORGANIC SUBSTANCES
(Production Volume >1,000,000 Lbs/Yr)

Production Range (Lbs/Yr)	Number of Substances	%	Total Production (Million Lbs/Yr)	%
$>10^{10}$	7	0.5	97,830	39.8
10^9 10^{10}	36	2.8	91,141	37.1
10^8 10^9	120	9.2	42,273	17.2
10^7 10^8	383	29.3	11,249	4.6
10^6 10^7	761	58.3	3,086	1.3
Total	1,307		245,580	

The volume distribution of 451 inorganic chemical substances (Table IV) represents 11.9% of the count of those chemicals produced in excess of one million pounds annually and 12.4% of the volume.

Table IV

VOLUME DISTRIBUTION OF INORGANIC SUBSTANCES
(Production Volume >1,000,000 Lbs/Yr)

Production Range (Lbs/Yr)	Number of Substances	%	Total Production (Million Lbs/Yr)	%
>10^{10}	17	3.8	348,731	69.3
10^9 10^{10}	35	7.8	112,934	22.4
10^8 10^9	89	19.7	36,443	7.2
10^7 10^8	133	29.5	4,375	0.9
10^6 10^7	178	39.2	775	0.2
	451		503,258	

Polymers and plastics (Table V) numbering 893 substances represent 23.5% of the count of substances produced in excess of one million pounds annually but only 3% of the volume. The largest volume substance, cellulose pulp, represents 64% of the total pounds produced in this category.

Table V

VOLUME DISTRIBUTION OF POLYMERS AND PLASTICS
(Production Volume >1,000,000 Lbs/Yr)

Production Range (Lbs/Yr)	Number of Materials	%	Total Production (Million Lbs/Yr)	%
>10^{10}	1	0.1	78,593	64.3
10^9 10^{10}	9	1.0	22,400	18.3
10^8 10^9	40	4.5	13,251	10.8
10^7 10^8	204	22.8	5,476	4.5
10^6 10^7	639	71.6	2,462	2.1
Total	893		122,182	

It is interesting to note that 380 separate chemical substances that make up the primary derivatives of petroleum represent 55.4% of all chemicals with production weight greater than one million pounds per year (Table VI).

Table VI

VOLUME DISTRIBUTION OF PRIMARY DERIVATIVES
OF PETROLEUM
(Production Volume >1,000,000 Lbs/Yr)

Production Range (Lbs/Yr)		Number of Materials	%	Total Production (Million Lbs/Yr)	%
>10^{11}		1	0.3	102,171	4.5
10^{10}	10^{11}	51	13.4	1,847,835	81.8
10^{9}	10^{10}	85	22.4	266,906	11.8
10^{8}	10^{9}	91	23.9	36,241	1.6
10^{7}	10^{8}	96	25.3	4,376	0.2
10^{6}	10^{7}	56	14.7	272	<0.1
Total		380		2,257,799	

The major substances with production volumes in millions of
pounds per year for organics, inorganics, polymers and plastics
and primary derivatives of petroleum are noted in Tables VII,
VIII, IX and X.

Table VII

MAJOR SUBSTANCES
Production Volume
(Million Lbs/Yr)

Organics

20,046	Propylene	5,944	Methanol
19,021	Ethylene	4,928	Styrene
13,837	Benzene	3,881	1,3-Butadiene
13,203	Urea	3,838	Acetic Acid
11,306	Butylene	3,793	Ethylene Glycol
10,383	Toluene	3,351	o-Xylene
10,036	Ethylene Dichloride	3,045	Cumene
8,603	Xylene	3,006	Ethylene Oxide
6,716	Ethyl Benzene	2,961	Formaldehyde
6,500	Vinyl Chloride		

Table VIII

MAJOR SUBSTANCES
Production Volume
(Million Lbs/Yr)

Inorganics

49,174	Sulfuric Acid	16,311	Nitric Acid
31,723	Calcium Oxide	15,127	Ammonium Nitrate
30,618	Ammonia	13,519	Sulfur
29,565	Sodium Hydroxide	13,136	Ammonium Phosphate (2:1)
29,413	Carbon Dioxide	13,068	Aluminum Oxide
19,169	Hydrogen	12,326	Calcium Hydroxide
18,498	Chlorine	11,583	Calcium Carbonate
18,125	Sodium Carbonate	10,304	Carbon Monoxide
17,073	Phosphoric Acids		

Table IX

MAJOR SUBSTANCES
Production Volume
(Million Lbs/Yr)

Polymers and Plastics

78,593	Cellulose Pulp
5,191	Polyvinyl Chloride
4,963	Polyethylene
3,622	Butadiene/Styrene Copolymer
2,107	Poly (Ethylene Terephthalate)
1,915	Polystyrene
1,245	Polypropylene
1,161	Urea-Formaldehyde Polymer
1,113	Phenol-Formaldehyde Polymer
1,082	Polybutadiene

Table X

MAJOR SUBSTANCES
Production Volume
(Million Lbs/Yr)

Primary Derivatives of Petroleum

102,171	Gas Oil (Middle)
99,525	Atmospheric Tower Residuum
93,700	Vacuum Residuum
82,381	Kerosine
80,856	Gas Oils, Heavy Vacuum
78,582	Naphtha, Heavy Straight Run
76,400	Gas Oils, Straight Run
73,960	Naphtha, Light Straight Run
71,230	Naphtha, Light Catalytic Reformed
69,140	Naphtha, Light Catalytic Cracked
68,135	Naphtha, Heavy Catalytic Reformed
66,310	Naphtha, Sweetened
65,875	Naphtha, Heavy Catalytic Cracked

Obviously, control decisions cannot be made on the basis of volume alone, but certainly the higher-volume materials deserve early scrutiny and consideration.

Current State of Knowledge of Health and Ecological Effects

Most of the commodity organic chemicals have rather complete data bases, although knowledge of certain effects may be missing. For the small-volume organic chemicals, health and environmental data bases are sometimes non-existent or limited to a knowledge of a few physical properties that may impact health and environmental effects.

Since many of the commercial chemicals have been in use for decades, it is apparent that major health and ecological effects should have been observed and reported in the literature if such effects occur under historical conditions of manufacturing and use.

Generally there appears to be consensus that testing to determine the health and ecological effects of many commercial chemicals should continue. However, the resources available for such testing are limited. In addressing this challenge of testing needs vs resource limits, it seems obvious that priorities for testing programs will have to be established.

The idea of prioritizing those substances which will be tested is not new. Every organization that has to consider the testing or assessment of chemicals has to prioritize its efforts. Some have done it formally and others informally. The U.S. EPA Interagency Testing Committee (ITC) has developed a procedure

which is used to focus attention on a smaller number of substances for consideration for priority testing. The Chemical
Manufacturers Association (CMA) has looked at these approaches
and has made a preliminary suggestion as to how materials could
be selected for testing.

As one reviews all the discussions on the need for testing
in order to obtain all the data necessary to make an appropriate
risk assessment of a substance, one gets the feeling that there
is so much to be done that there must be very little going on or
completed. This is not the case at all. There are a large
number of organizations, both governmental and private, which are
engaged in testing of existing chemical substances.

Since there is no concerted effort to coordinate testing
underway in various laboratories of the world, it might be felt
that this lack of coordination may cause some important substances to be missed and thereby present the possibility of harm
being done to people or the environment. In order to determine
if this were so, we undertook an examination of the top 50 chemicals produced in the U.S. This is a list of commercial chemicals
selected by *CHEMICAL AND ENGINEERING NEWS* as those materials with
the largest production. It should be noted that these are dis-
tinct substances, not mixtures or petroleum products (Table XI).

Without trying to carry out an exhaustive search, we accumu-
lated information on the physical, chemical, toxicological and
ecological properties of these 50 substances. The amount of data
available varied for each chemical, as would be expected. We
found that there were physical and chemical properties on all 50.
In the case of data on human health and mammalian toxicology,
there were only five substances for which there was no information, and these were in every case substances that are well known
and would not be expected to present a problem. In the case of
ecotoxicology there was less, but there still are data for all
except for eight substances: ammonium sulfate, carbon dioxide,
sodium carbonate, nitrogen, oxygen, sulfuric acid, terephthalic
acid and water glass.

Another approach that we took towards the top 50 was to
examine what the ITC had done with these substances. We found
that of the 50 substances only 18 survived the screening process
used by the ITC to narrow down the total number of substances
examined. Each of these 18 was scored, using the ITC scoring
procedures, and were then considered by the committee. Four have
been recommended for priority testing: ethylene oxide, propylene
oxide, toluene and xylene.

We also reviewed the testing programs of the Chemical
Industry Institute of Toxicology (CIIT) and CMA as they pertain
to the top 50 chemicals. We found that there were 21 substances
which are either being tested or are being considered for testing.
An examination of the testing programs of these two organizations
also shows that there are testing programs underway or planned
for another 35 of the major commercial substances which are not

Table XI SUMMARY OF DATA AVAILABLE AND TESTING PLANNED ON U.S. "TOP 50" CHEMICALS

Chemical Substances	Data Available*			Testing Planned or Underway	Scored by ITC (R=Recommended)
	Physical Properties	Human Health Mammalian Toxicology	Ecotoxicology		
Acetic Acid	X	X	X		
Acetic Anhydride	X	X	X		
Acetone	X	X	X	X	
Acrylonitrile	X	X	X	X	
Adipic Acid	X	X	X		
Aluminum Sulfate	X	X	X		
Ammonia	X	X	X	X	
Ammonium Nitrate	X		X		
Ammonium Sulfate	X	X			
Benzene	X	X	X	X	X
1,3-Butadiene	X	X	X		X
Calcium Chloride	X	X	X		
Calcium Oxide	X	X	X		
Carbon Black	X	X	X		X
Carbon Dioxide	X	X			
Chlorine	X	X	X	X	
Cumene	X	X	X		X
Cyclohexane	X	X	X		X
1,2-Dichloroethane	X	X	X	X	X
Ethanol	X	X	X		
Ethyl Benzene	X	X	X		X
Ethylene	X		X	X	X
Ethylene Glycol	X	X	X	X	X
Ethylene Oxide	X	X	X	X	R

Table XI (Cont'd.)

Chemical Substances	Data Available* Physical Properties	Data Available* Human Health Mammalian Toxicology	Data Available* Ecotoxicology	Testing Planned or Underway	Scored by ITC (R=Recommended)
Formaldehyde	X	X	X	X	X
Hydrochloric Acid	X	X	X		
Methanol	X	X	X	X	
Nitric Acid	X	X	X		
Nitrogen	X				
Oxygen	X				
Phenol	X	X	X	X	
Phosphoric Acid	X	X	X		
Propanol	X	X	X		
Propylene	X		X		X
Propylene Oxide	X	X	X	X	R
Sodium Carbonate	X	X			
Sodium Hydroxide	X	X	X		
Sodium Sulfate	X	X	X		
Sodium Tripolyphosphate	X	X	X		
Styrene	X	X	X	X	X
Sulfuric Acid	X	X			
Terephthalic Acid	X	X		X	X
Titanium Oxide	X	X	X	X	X
Toluene	X	X	X		R
Urea	X	X	X	X	
Vinyl Acetate	X	X	X	X	
Vinyl Chloride	X	X	X	X	
Water Glass	X	X			
p-Xylene	X	X	X	X	R
Xylene	X	X	X	X	X

*Results of preliminary searches. It is not exhaustive and does not reflect proprietary data.

the top 50 volume chemicals in the U.S. There is no reason to
believe that a survey of the literature would not yield similar
results as those discussed for the top 50 for the other major
commercial chemicals.

It is apparent from this discussion that if one narrows down
the large universe of commercial chemicals to those produced in
sufficient quantities, such that they might be expected to pose a
possible health or unwarranted threat, there is a significant amount of information known about these substances or
studies are being considered for a number of them. This is not
to say that there is no need for continued testing of existing
commercial substances, but the picture is not as bleak as some
would want us to believe.

There are considerable resources being devoted to the testing of existing substances throughout the world. This testing is
being concentrated on those substances which would be expected
to pose risk to health or the environment.

Priority Setting

Typically, the dossier of a chemical will show gaps in the
data. These gaps can be considered as voids that may need to be
filled by testing. Simplistically, among the groups of chemicals
under consideration, the chemicals with the greatest number of
gaps and the greatest exposure potential would receive priority
consideration.

Realistically, however, priority-setting is more complex.
Expert judgment, utilizing analogy and experience, is necessary
to assess qualitatively the most sensitive toxicological or
ecological effect likely to be of concern, and for this effect(s)
it is necessary to estimate the amount of uncertainty in the risk
on the basis of existing data. An estimate then needs to be made
which will determine the degree to which the uncertainty will be
reduced by further testing.

A panel of experts would be expected to take into consideration the accumulation of experience with a given substance,
including past exposure as well as current steps being taken, to
control the exposure to the substance. A substance which has
been produced in significant quantities over a long period of
time with no known adverse effects would be of less concern than
a substance whose production is rising rapidly and for which
there is little experience. If the exposure to a substance has
been reduced through changes in production or use patterns, that
substance should also receive less attention than one over which
no control is exercised. When available, epidemiological data
should be considered along with animal data.

The expert panel can bridge the information gaps of concern
for a given chemical and provide a qualitative ranking of those
risk uncertainties which are to be clarified by testing. This
case-by-case approach will be directed by the basic interests and

mission of the institution conducting the prioritizing of the substances. Thus, we see here an approach for selection of substances based on a well-defined procedure for narrowing the overall universe followed by case-by-case consideration by appropriate experts.

In the U.S. it has become apparent that the number of materials which need to be considered for priority testing is such that, even after the screening approach described is applied and case-by-case selections are made, it will require our combined resources the next five to 10 years to adequately fill the data gaps so identified.

Such prioritization is needed to assure that scarce testing resources are focused on those materials of greatest concern.

RECEIVED September 1, 1982

Overview After Five Years

E. HAMILTON HURST

Nalco Chemical Company, Oak Brook, IL 60521

The Toxic Substances Control Act (TSCA)
was signed into law on October 11, 1976
and became effective on January 1, 1977.
During its formative first years, there
have been relatively few regulations
finalized under the potentially wide
scope of the law. Even so, TSCA has
had a perceptible and important impact
on the chemical industry and the way it
operates. This overview of progress
under the TSCA law, will highlight and
differentiate portions of the law where
regulations are finalized, pending or
awaiting administrative development.

On January 1, 1977, the chemical industry truly
became a regulated industry. The environmental laws
up until that time had covered some chemicals, but had
been media oriented. That is -- they were concerned
about certain chemicals that escaped as emissions or
pollutants to various media - the air, our water,
contaminated our food or entered the workplace. TSCA
changed that direction. It was designed to regulate
commerce on chemical substances. TSCA potentially
applies to all chemicals manufactured, processed,
distributed or used in the U.S. except those chemicals
already regulated under certain other federal laws.
TSCA affects not only the chemical industry itself,
but the many other industries whose products are
chemical in nature. This includes most all industrial
products.
    The origin of TSCA is attributed to the CEQ
(Council on Environmental Quality) report "Toxic
Substances" prepared on April, 1971. This report
focused on certain metals and their compounds and

synthetic organic chemicals as toxic substances for
which insufficient information was known and which,
thereby, may significantly threaten man and his
environment. From this initiative, legislation was
introduced into Congress in 1971, but Congress,
through 1971 and 1972, could not agree on a common
legislative thrust and language. Time ran out in the
92nd Congress before a law was enacted.

This respite was short lived. The next Congress,
the 93rd, tried again in 1973 and it too could not
agree on a suitable bill. For nearly four years,
Congress searched for a legislative approach which
would afford adequate protection to human health and
the environment while preserving an innovative and
dynamic chemical industry. It was left to the 94th
Congress to reach agreement, which they did in the
Fall of 1976 and the legislation was signed into law
by President Ford in October, 1976.

So, after 6 years of debate, TSCA was born. This
was an important six years. Many of the environmental
laws of our country were enacted during the 1960's and
early 70's. TSCA was to be the "cap" on all of the
laws - filling all the gaps that existed between the
previous laws. It was also designed to put in place,
a law to regulate all chemicals in commerce which may
present an unreasonable risk in any part of the
chemical's life cycle. Any part of the life cycle can
be regulated from R&D through production, distribution
and disposal.

But something happened in the 6 years it took to
pass TSCA through Congress. The country awoke to find
that some of the environmental laws enacted in the
60's and early 70's had a significant impact on the
economy and on jobs. The laws as written did not
require consideration of economic consequences. The
Clean Air Act and the Clean Water Act were technology-
forcing laws that required the use of feasible technological controls without concern for economic impact.
Similarly our Occupational Safety and Health Act was
also, in the setting of standards, a technology-
forcing law. The thrust of these laws were technology
forcing since the safe level for exposure was assumed
to be synonymous with the lowest level achieveable.

In contrast, the Food Additive Amendments of 1958
to the Food, Drug and Cosmetic Act presented a different type of law, a "zero-risk" law. Under the thrust
of the zero-risk law, it said "if a food additive is a
carcinogen, we cannot permit any exposure since we
cannot live with any such risk." Risk is the evaluation of severity of toxicity or hazardous properties

as related to the exposure of humans or the environment to the hazardous property. Under a zero-risk
law, the regulation focuses on the hazards. If there
is a hazard, there can be no acceptable exposure, no
matter how small or insignificant. Under a zero-risk
law, if the chemical presents a hazard, it is an
unacceptable product.

In the late 1970's, the country's mood changed to
accept not just some risk, but to include in the
decision equation, specific consideration for economic
factors, social benefits and impact on jobs. And so,
the Clean Air Act was amended in 1977, changing the
thrust of the law to include economic and practical
considerations in carrying out the mandate of that
law. And so too, the TSCA law that finally passed in
late 1976, was a "balancing law," the first really
true balancing law to be passed by Congress. It contains as part of the policy and intent statement
(Section 2) of the Act, the following:

Section 2 (b) Policy - It is the policy of the
U.S. that:

(1) adequate data should be developed with respect to the effect of chemical substances and mixtures on health and the environment and that the development of such data should be the responsibility of
those who manufacture and those who process such
chemical substances and mixtures;

(2) adequate authority should exist to regulate
chemical substances and mixtures which present an unreasonable risk of injury to health or the environment,
and to take action with respect to chemical substances
and mixtures which are imminent hazards; and

(3) authority over chemical substances and mixtures should be exercised in such a manner as not to
impede unduly or create unnecessary economic barriers
to technological innovation while fulfilling the
primary purpose of this Act to assure that such innovation and commerce in such chemical substances and
mixtures do not present an unreasonable risk of injury
to health or the environment.

Section 2 (c) INTENT OF CONGRESS -- It is the
intent of Congress that the Administrator shall carry
out this Act in a reasonable and prudent manner, and
that the Administrator shall consider the environmental, economic, and social impact of any action the
Administrator takes or proposes to take under this Act.

These statements of policy and intent are important inclusions in the TSCA law. Congress recognized
the various approaches that had been taken in the past
were not appropriate. It decided that a different

approach was needed for regulating an industry whose
very purpose in society requires that it and its people deal with chemical products, some of which have
hazardous properties.

So in TSCA, we have a "balancing-type law" where-
in the Administrator is required to consider not just
the risks associated with a chemical, but also whether
it is an unreasonable risk in light of the benefits
associated with the chemical. For the Administrator
to regulate a chemical substance or mixture under
Section 6 of TSCA, the law requires that "the Administrator shall consider...

(A) the effects of such a substance or mixture on
health and the magnitude of the exposure of human
beings to such substance or mixture,

(B) the effects of such a substance or mixture on
the environment and the magnitude of the exposure of
the environment to such substance or mixture,

(C) the benefits of such a substance or mixture
for various uses and the availability of substitutes
for such uses and,

(D) the reasonably ascertainable economic consequences of the rule, after consideration of the effect
on the national economy, small business, technological
innovation, the environment and public health."

It is this background of the legislative intent
of our Congress that makes TSCA a unique law. For the
most part, the burden for making the necessary "unreasonable risk" finding rests with the government,
and the law provides for information gathering
(Section 8) from industry or others to help make this
finding. Since each chemical has certain properties
and, in its use, has certain exposures, potential
risks, and certain benefits to society, each evaluation of the unreasonableness of the risks will depend
on its own unique set of factors. Congress did not
attempt to define unreasonable risk. Congress did not
leave the agency an easy job. TSCA demands a balanced
approach and it is this approach, with all the complexities it may present, that is the essence of TSCA.
It is the only way such a law in our country can
effectively regulate chemicls without severely affecting innovation in the chemical industry.

With this background then, let us shift now to a
discussion of principal provisions of the law, what
timetables were built into the law and how far along
the road we have come.

The basic thrust of TSCA is to control unreasonable risks, not all risks -- but unreasonable risks.
To accomplish this, TSCA proposes that adequate data

be developed with respect to the effect of chemical
substances and mixtures on health and the environment
and that authority exists to regulate those substances
and mixtures which present an unreasonable risk.
This thrust is carried out in several ways
throughout the Act in the principle sections:
Section 4 - authorizes the EPA to promulgate
rules to require manufacturers and/or processors to
test specific chemical substances or mixtures in order
to evaluate their human health or environmental
effects. This can be required when chemicals are suspected of being hazardous or when produced in substantial quantities which may present significant exposure
to humans or the environment.
Section 4 has three other significant provisions:
(1) It allows the EPA to prescribe the procedures
or methodology used to conduct the testing.
(2) Testing costs are to be borne by the manufacturers and/or processors and if these firms cannot
agree on allocation of testing, the EPA can establish
the allocation of costs.
(3) Section 4 sets up the Interagency Testing
Committee which can designate chemicals for priority
testing. The EPA must respond to such designations
by issuing a testing rule within 12 months of the
designation or indicate why such testing is unnecessary.
Section 5 - empowers the EPA to screen new
chemical substances and existing chemical substances
to be manufactured for significant new uses before
manufacturing starts. The industry is required to
submit a notice describing the chemical and its use 90
days prior to manufacturing. The Agency has broad
powers to limit manufacture or use if the EPA concludes that the chemical may present an unreasonable
risk and significant unanswered questions exist.
Section 5 also permits exemptions for R&D purposes,
application for test marketing exemptions, and for
those chemicals which present no unreasonable risk.
Section 6 - authorizes the EPA upon determination
of an unreasonable risk to, by rule, ban the chemical,
prohibit or limit certain uses or require labeling.
In making the determination of unreasonable risk, the
Agency must balance the risk against the economic and
social disadvantages of eliminating or restricting the
availability of the chemical. Section 6 also requires
the Agency to regulate PCB's (Polychlorinated Biphenyl's), the only chemical specifically targeted by
the law itself.
Section 7 - authorizes the EPA to move directly
to Federal District Court against chemical substances
or mixtures which are imminent hazards.

Section 8 - enables the EPA to require recordkeeping and reporting of information. Part of this
authority includes the direction for the Agency to
compile an inventory of all chemical substances in
commerce. The basic information gathering activities
are: Section 8(a) - reporting of production, use,
exposure and disposal information on specific chemicals; Section 8(c) - recordkeeping of significant
adverse reactions; Section 8(d) - submission of lists
and copies of health and safety studies; Section 8(e)
- notification of substantial health or environmental
risks.

To complete the comprehensive coverage of TSCA,
the following Sections of TSCA also deserve your
attention:

Section 9 - defines the relationship between TSCA
and other Federal laws. In general, this section
seeks to avoid overlapping or duplicative activities
by the various Federal agencies with authority over
chemicals.

Section 10 - authorizes the EPA to conduct research and to develop an effective system to collect
and disseminate data submitted under this Act to
others.

Section 11 - authorizes the EPA to conduct inspections and to issue subpoenas to collect information for enforcement of the Act.

Section 12 - describes the EPA's authority regarding chemical substances and mixtures manufactured
for export.

Section 14 - outlines the responsibility of the
Agency to protect trade secret and confidential business information it gathers in the process of implementing the Act.

Sections 15 and 16 - define prohibited acts under
TSCA and prescribe penalties for violating the Act.

Sections 20 and 21 - authorize citizen action to
compel compliance with the regulatory requirements of
the Act.

These sections are all inter-related. They were
intended to be that way. It should not be and is not
the intent of the law that the Agency would administer
each section separately. For example, Section 8
should not be used to gather information for information sake only. Section 8 is designed to help focus
on those chemicals selected for consideration under a
testing rule (Section 4) or regulation (Section 6).

In the law itself, several sections are self-implementing and others have specified timetables.

For example:
  Section 8(e) is a reporting responsibility which
went into effect January 1, 1977 (the date the law was
effective). This requires any manufacturer, processor
or distributor who learns of information which supports a conclusion that a substance or mixture presents a substantial risk, must report this to the
Agency immediately.
  This was the first part of TSCA which impacted
the industry. The responsibility of 8(e) reporting has
been with the industry since Day One of TSCA. While
not required to, the Agency did publish a guidance
document in March of 1978 to assist industry in understanding its responsibilities. To date, some 400
notices have been submitted to the Agency. While the
Agency has followed up with the notifiers, no other
action has been taken by the Agency to date on the
chemicals involved.
  A second major provision of TSCA which is self-
implementing, is the requirement for a pre-manufacturing notice on new chemical substances. This requirement was triggered by the publication of the Initial
TSCA Inventory in May of 1979. By law, the PMN program started 30 days after the inventory of chemical
substances was published by the EPA. Starting on
July 1, 1979, the PMN program has proceeded very
successfully. As of February, 1982, some 1,100 PMN's
have been reviewed through the PMN System. PMN's are
being submitted at the rate of 700/year currently.
  The remaining sections of TSCA require specific
rulemaking or action by the Agency to implement the
various provisions of the Act. The Agency has not
been idle and has proposed numerous rules and procedures.
  The Agency under the Carter Administration, however, tended to propose regulations which were judged
by many to be beyond the statutory requirements of the
law and in many cases unworkable. Of primary concern
has been the adverse effect of PMN reporting regulations on innovation within the industry and the lack
of an identifiable purpose and reasonableness in the
various reporting rules proposed. The thrust of
Agency action seemed to be that each section of TSCA
should be implemented on its own merits without consideration for the overall inter-related actions and
purposes of TSCA.
  Thus, much of what the Agency has proposed, has
been the subject of industry concern. Many in industry have submitted comments to the Agency expressing
their objection to the approach taken by the Agency.

At the same time, others such as the environmental groups have argued that TSCA is not working and that EPA has not fulfilled their obligation to implement the law. They have been particularly critical of missed deadlines by the Agency, that only a few final rules developed and the failure to regulate chemicals under TSCA. It would appear that their measurement of success of TSCA would be -- How many rules are written? or -- How many chemicals are banned? This does not seem to be a clear-thinking approach.

If your measurement of effectiveness of the law is what action has been taken, then here is how your scoreboard looks:

EPA Actions Under Section 6

10/15/78 - Ban use of CFC's (Chlorofluorocarbons) as aerosol propellants in non-essential uses.

5/31/79 - Final rule on PCB struck down by U.S. Court 10/31/80. Court ordered interim program 3/10/81.

3/11/80 - Final rule on disposal of wastes containing 2,3,7,8 Tetrachlorodibenzo-p-dioxin.

9/17/80 - Proposed rule on asbestos in school buildings.

10/7/80 - ANPR restricting production of CFC's.

EPA Actions Under Section 5

11/26/80 - Proposed significant new use rule (SNUR) n-Methanesulfonyl-p-toluene sulfonamide.

EPA Actions Under Section 4 (Testing)

Proposed Test Rules. 7/18/80 - Chloromethane.
7/18/80 - Chlorinated Benzenes.
6/5/81 - Nitrobenzene
6/5/81 - 1,1,1 Trichloroethane.
6/5/81 - Dichloromethane

Notice of Decisions Not to Propose Testing Rules

Acrylamide (health effects).
Polychlorinated Terphenyls.
Chlorinated Napthalenes.
Benzidine dyes.
0-toluidine dyes.
Dianisidine dyes.

To those of us who have worked closely with TSCA through the past 5 years, the Act is working, and has

achieved many of the goals established. This impact
is not necessarily measurable by the yardsticks such
as numbers of rules or regulatory actions. However,
the impact is measurable. It can also be demonstrated
by the conduct of the industry responding to the public and political environment which has led to the
passage of TSCA and other environmental control laws.
Today, responsible companies in the industry have
initiated internal mechanisms which have produced the
results intended by law.

An illustration of this has been the response of
industry to the 8(e) notice requirements of the law.
Most companies have established an internal communications system to collect potential 8(e) information,
selected personnel to evaluate the information
gathered, and have developed an expertise in handling
this type of hazard reporting. Typically, the 8(e)
notifier takes appropriate action or response on his
own initiative to control or alleviate the risks involved. It has been commonplace for the notifier to
advise his customers as well as his employees for the
chemical involved of the information contained in the
notice.

The list of 8(e) notices as they have been published or otherwise made available by the EPA or trade
press are followed closely by health and safety departments of companies in all types of industries.
Industry in general has developed methods to not just
acquire and evaluate information, but also take responsive action.

Part of the ability of industry to respond to
information such as 8(e) notices has been the result
of development of health and safety units throughout
industry. While some companies have had identifiable
safety departments for many years, in many companies
today, the organizational chart shows a health and
safety department usually reporting at a high level of
management. Obviously, each company has its own system or organization reflecting its own management style
but, the point is -- companies respond in some way
organizationally to fill the need.

I suggest that these steps would have occurred
with or without TSCA or other laws being in place. I
personally believe companies do respond to the needs
expressed by the public and that the "marketplace" --
the public's perception of what the responsible
company should do -- has significant impact.

Without question, these illustrations demonstrate
the fact that companies will respond to the thrusts of
laws with or without the promulgation of specific

regulations. In the law, Section 8(e) is only one
short paragraph (8 lines long) that establishes the
thrust of the law and goal to be accomplished. That
is all we in industry should need -- tell us what the
goal is and let us determine how we can best achieve
it. We do not need specific instructions as to how we
should carry out our responsibility to comply with the
law.

     In my opinion, the debate over the progress of
TSCA thus far has found root in part from the lack of
perception of what TSCA law really is. The significant changes in TSCA that occurred during its six-year
gestation period may not be fully recognized nor accepted by some. It is the first truly "balancing law"
the Agency has had to administer. In some respects,
the expectation of some that TSCA should be rapidly
regulating chemicals does not appreciate all of the
requirements of TSCA and its policy statements.

     Congress in writing TSCA Section 2 said the Con-
gress finds that -- among the many chemical substances
and mixtures which are constantly being developed and
produced, there are some -- whose manufacturing may
present an unreasonable risk--. This and other provisions of TSCA establish that the Congress, in reaching a decision on legislative approach, recognized
that not all chemicals would present unreasonable
risks. It anticipated that the Agency would have to
seek out which chemicals may present unreasonable risk,
establish priorities and proceed, through rulemaking,
to regulate. In the first 4 years of TSCA, the EPA
did not do this job well. The Agency is just now developing a program of exemptions from the PMN process
for those chemicals which do not pose unreasonable
risks.

     To help set priorities for testing, the EPA has
available, the assistance of the Interagency Testing
Committee (ITC). The ITC was established by the
Agency as directed by law and it came forward with its
initial list of designated chemicals in October, 1977,
meeting the time scheduled by the law. Approximately
every six months thereafter, the ITC has submitted
additional names to the Agency to add to the list. In
all, some 9 lists including some 49 chemicals or
chemical categories have been designated.

     One of the criticisms of the Agency has been that
they have failed to take action within the statutory
12-month period on the ITC list of chemicals. This
resulted in a suit by the Natural Resources Defense
Council (NRDC) in May, 1979. The District Court
ordered the EPA to prepare a plan for complying with

the law. The EPA is now under court order to adhere
to a schedule agreed upon for those chemicals in the
first 6 lists submitted by the ITC.

Under the present EPA administration, there is
every indication that TSCA implementation will speed
up. The recently available draft of TSCA "Priorities
for OTS Operation" prepared by Don Clay and his management team, suggests the Agency will direct their efforts
toward two broad areas:

    1)   New Chemical Program
    2)   Existing Chemicals Program

The New Chemical Program Has Four Major Parts:

    1)   Review of PMN's submitted
    2)   Finalizing PMN requirements
    3)   Establishing PMN exemptions
    4)   Follow-up program for new chemicals

The Existing Chemical Program Has Five Major Parts:

    1)   Focus on reduction of unreasonable risks
    2)   Emphasize voluntary control by industry and
the public
    3)   Concentrating evaluation and control efforts
on specific problems
    4)   Directing effort at problems identified
through specific TSCA mechanisms
    5)   Exchange technical information with industry,
labor and others on specific problems

The report outlines in more detail the overall
approach planned by the EPA for implementing TSCA. As
a general comment, it would appear the Agency will
focus its activity on those chemicals which are perceived to present major risks and will use the entire
list of options open to them under TSCA to effectively
evaluate the risk and take appropriate action. Their
thrust will also apparently be to encourage voluntary
action on the part of industry to control those situations which may present unreasonable risks. This will
be particularly true for the testing program for
chemicals including those on the ITC lists. Clearly
the Agency expects the ITC list to serve as a priority
for them and I believe we can expect to see action
taken in the future on ITC-listed chemicals within the
statutory time limit.

The overall outline of action indicated in the
100-Day Report is encouraging. The Agency seems ready
to proceed with the implementation of TSCA in an
orderly manner focused only on those particular

chemicals which fit the definition in Section 2 --
"some whose manufacture, processing, or distributing
in commerce, use or disposal may present an unreasonable risk." Five years is a long time. However, when
implementing a law as comprehensive as TSCA on an
industry as complex as the chemical industry, it is
far too short. Significantly, the Agency has accomplished two major actions: 1) the creation of the
TSCA Inventory and, 2) the review of over 1,100 PMN's.
Both tasks are noteworthy.

The inventory of some 55,000-plus chemicals was
a major task. This document is critical to the working
of TSCA and is being "copied" by others such as the
European Economic Community in establishing their
equivalent of TSCA for the 10-member countries of the
European Common Market. The inventory in the U.S. has
become a valuable tool not only for the Agency as it
works with TSCA, but also for the chemical industry.
It ranks in equal importance with all of the other
major reference books used by the industry in its R&D
activity and commercialization plans.

The PMN program has also been highly informative.
With an experience factor developed by reviewing over
1,100 PMN's, a review of the data shows that the Agency
has found reason to question only a very few chemicals
as presenting unreasonable risks. While we in industry
might have believed that -- there is nothing like --
"seeing it -- to believe it." This experience alone
has been an important lesson learned by the Agency
which should have dramatic influence on what the Agency
does from here on implementing TSCA. We hope that
other countries in the European Economic Community and
the greater number in the Organization for Economic
Cooperation and Development will use this experience in
putting their somewhat different systems into operation
to assess the risks of new chemicals.

The PMN program also demonstrates one other very
significant fact. Most of the "new chemicals" coming
through the PMN system are merely modifications of an
existing type of chemistry. This fact has been helpful
to the Agency since experience with the family of
chemistry is helpful in making their evaluation and
assessment of risks.

But this fact also says one other thing -- our
industry's R&D effort is not producing "new chemistry."
We continue to massage the old chemistry. Does this
reflect the impact of the PMN regulation on innovation
and invention? Are we not exploring new fields in R&D
because the "new" fields of chemistry may be difficult

to pursue with the uncertainty of the PMN process facing us? Or is it because our productivity in research is falling for reasons other than regulatory pressure? Whatever it is, it is clear that innovation has been adversely affected. The PMN process of TSCA as implemented thus far, has played a significant role in causing this, but it is the challenge for R&D management in our industry to take the steps necessary to create a climate in our research activities for innovation. The challenge is also present for the Agency to assure that their rules and regulations place only reasonable and prudent burdens on the industry as it invests in the new frontiers of the chemical world.

So the industry has completed 5 years of life under TSCA. We are still in business. But we have changed. Some of the changes come from responding to the law. Some of our responses are to the "spirit of TSCA" -- the public demands for responsive action that created the political atmosphere in which TSCA was originally conceived. From the EPA, we have seen several false starts in getting their act together. This should not be too surprising. They started from scratch in 1977 with only a handful of people. Today the Office of Toxic Substances is 600 strong. In fairness to the Agency, it takes time to hire and train the people to do an effective job with a new "balancing-type" law.

Where to from here? -- it is not difficult to predict. We will see TSCA implemented -- hopefully in a reasonable and timely manner. The Agency has learned a lot about our industry, about our new chemicals and about how to implement the law effectively. Good communication between the Agency and the industry will be a "must" if the law is to work efficiently. While we may come to the discussion table from two different backgrounds and viewpoints, the only way to leave the table is with joint understanding of what truly needs to be done to manage the risks of interest in a reasonable and responsive way. It is in our overall best interest to see this done. It is simply good business. It is also good government! And it will require our best efforts to present our viewpoint positively, persuasively and effectively.

Literature Cited

1.   "Legislative History of the Toxic Substances Control Act," prepared by the Environment and Natural Resources Policy Division of the Library of Congress for the House Committee on Interstate and Foreign Commerce.

2.   Chemical Manufacturers Association; "The First
Four Years of the Toxic Substances Control Act."
3.   Clay,Don, Director-Office Toxic Substances, "Priorities for
OTS Operation (100-Day Report)," February, 1982.
4.   Conference Report 95-564, 95th Congress, Clean Air
Amendments of 1977, Section 313, p 109-110.
5.   Toxic Substances Control Act, U.S. Public Law
94-469, October 11, 1976.
6.   Occupational Safety and Health Act, Public Law
91-596, December 29, 1970.
7.   Federal Water Pollution Control Act, Public Law
92-500, October 18, 1972.
8.   Food and Drug Act, Public Law 85-929, September 6,
1958.

RECEIVED September 29, 1982

Initiatives of Chemical Industry to Modify TSCA Regulations

STACIA DAVIS LE BLANC

Chemical Manufacturers Association, Washington, DC 20037

By way of brief background, I would like to summarize how CMA has organized itself to deal with TSCA, and describe some of the basic themes which we have followed for virtually all of the issues raised by the Act.

First of all, CMA conducts what are described as "Advocacy Activities" -- a term which embraces all of our communications with Congress, the regulatory agencies, the courts, and the administration. These activities are multidisciplinary. Whenever an advocacy issue is identified, that matter is staffed by various disciplines in CMA including Technical, Legal, Government Relations and Public Relations where appropriate.

CMA institutionalized the management of toxic substances control matters in the Chemical Regulations Advisory Committee (CRAC). This committee has a regime of task groups staffed by member company executives which deal with the various sections of TSCA. For instance, there is a testing committee for Section 4 matters, a reporting task group for Section 8 and related matters, and so on. The coordination of these task groups is managed by CRAC.

CRAC is the basic client organization within CMA. It consists of 15 company representatives, one-third of which are replaced each year by new representatives. This standing technical committee is served by CMA staff, outside consultants and outside counsel, as appropriate. While the various sections of the Toxic Substances Control Act are interrelated and interdependent this section-by-section task group structure has so far proved extremely effective in dealing with the various proposals which EPA has issued. Our key organizational theme, therefore, has been to utilize a multidisciplinary team concept to

0097-6156/83/0213-0095$06.00/0
© 1983 American Chemical Society

serve a specific client organization within CMA, which
is responsible for addressing all TSCA related issues.

The general CMA philosophy in addressing TSCA
issues over the preceding months and years has been
very much a reflection of how EPA has addressed the
Toxic Substances Control Act. Unfortunately, under the
previous Administration, the philosophy of those within
the Agency was to propose an implementation of the
statute which in CMA's opinion exceeded the statutory
limits of the provisions of TSCA. We refer to this as
"statute busting." There may come a time in the
history of the administration of a statute, when
extending the limits of authority which Congress has
bestowed upon an agency is appropriate. It clearly
does not seem to be appropriate, however, for a statute
in which the government has had little or no regulatory
experience. In any event, this is clearly the case
with TSCA. Not only did the previous Administration
try to expand its authority under TSCA, but the Agency
further complicated the implementation of TSCA by
creating an extremely compartmentalized organizational
structure. This resulted in Section 4 people not
talking to Section 5 people, who in turn did not talk
to Section 6 or Section 8 people -- and so on. What
resulted was an ineffectively structured and
unnecessarily expansionistic attitude toward TSCA.

This Agency approach, necessarily, had a great
deal to do with how our industry responded to the
Agency's proposals. First of all, a major portion of
our comments had to define the rational limits and
scope of the Agency's statutory authority. Some people
thought we were excessively legalistic. We did not
apologize, however, because it appeared clear to us
that the Agency was going beyond the law in both fact
and spirit.

Next in our response to EPA's proposals was to
present a detailed critique of the particular proposal
which the Agency was bringing forth, and in addition --
and this is most important -- to offer a constructive,
rational alternative to that proposal, which not only
met the goals of the Act, but was within the limits and
the scope prescribed by Congress.

This three-part philosophy of our responses --
legal evaluations, policy critiques and technical
evaluations based on the need for "good science" at all
regulatory stages and presentation of constructive
alternatives, led to other operating philosophies.

In general, because the Act itself is
interrelated, we had to address virtually every TSCA
proposal by the Agency. We left product specific

proposals to the affected groups of manufacturers to
raise matters specific to that particular product. We
did take the opportunity to address generic issues in
those proposals, however, in an effort to protect the
larger interests of our industry in future rulemaking
activities.

Another practical operating philosophy which we
followed was to carefully and extensively communicate
with other trade associations and business groups to
assure the positions we took were not inappropriately
inconsistent with positions taken by those
organizations. A considerable amount of time and
paperwork was devoted to this interindustry
communications effort.

Another theme which we constantly reiterated in
all of our communications with EPA on TSCA, and indeed
in all other advocacy matters, was the need for
regulatory action to be based upon adequate scientific
data and defensible conclusions. Historically, the
Agency had been too quick to impose requirements on the
basis of incomplete, controversial or ambiguous
scientific evidence.

We also take the approach in our presentations to
the government -- in TSCA as well as in other advocacy
issues -- to request less reliance on formal
requirements and more emphasis on voluntary action and
informal negotiation to achieve TSCA's objectives.
Inflexible regulations are time-consuming, expensive
and cumbersome to implement. Enforceable performance
oriented standards can be designed. Unnecessary
disputes over the legality of EPA's actions might be
prevented, and needed regulations can be put in place
faster.

While several EPA initiatives under TSCA have
become final, many more are in the formative stages.
Under Sections 4, 5, 6, 8 and 12 of TSCA, EPA has
developed a large number of proposals, some of which
await final action. If implemented in a reasonable
manner, many of EPA's pending proposals can provide a
viable basis for implementing TSCA. I want to
emphasize, in fact, that CMA believes that TSCA as
written is workable. The following discussion will
demonstrate how industry has worked with the EPA in
applying the major provisions of TSCA in a manner that
is reasonable, cost-effective and consistent with the
original intent of Congress.

Section 4 of TSCA authorizes EPA to require
manufacturers or processors to test specified existing
chemical substances when available data and experience
are insufficient to evaluate their health and

environmental effects. The EPA may, under certain
circumstances, prescribe the methodology used to
conduct the testing to assure reliable results.
Testing costs are borne by manufacturers or
processors.

CMA has stressed the need for EPA to establish an
adequate data base on prospective chemicals before
publishing a proposed test rule. The Agency needs to
be fully informed about the chemicals under
consideration for testing. The Interagency Testing
Committee (ITC) which recommends chemicals for testing
to the EPA, needs to develop a more complete profile of
testing candidates. This has been accomplished to some
extent by having the ITC work more closely with
industry at the early stages of an assessment of a test
rule.

This approach is a significant improvement to some
of the initial proposals. In the past, EPA devoted
excessive time and effort to overly ambitious testing
approaches that were questionable on legal as well as
scientific grounds. For example, ITC had recommended
several broad "categories" of chemicals for EPA
consideration·for testing that raised issues of
defining the members of the category, and allocating
costs among the companies who make the chemicals within
that defined category. CMA has recommended that ITC
avoid broad designations and list the specific
chemicals if appropriate.

EPA's testing program was initially excessively
legalistic and formalized. Recently, however, the
Agency has entered into dialogue with industry. We
have encouraged this development generally, and
specifically promoted consideration of EPA/industry
negotiated testing programs. By ensuring that industry
is aware of EPA's testing priorities and by
communicating with industry representatives at an early
stage of the test rule development process, EPA can
promote voluntary testing of chemicals. Recently, much
of the needed testing is being conducted by industry as
a result of voluntary negotiations between EPA and the
affected manufacturers. As a result, chemicals of
priority concern are being further evaluated more
efficiently than if EPA initiated a full rulemaking
proceeding.

CMA has also urged greater reliance on informed
scientific judgment of the test sponsor. This is in
line with our general theme that regulatory decisions
should be made on the basis of "good science." EPA had
expressed great interest in new and controversial areas
of testing, such as neurotoxicity and mutagenicity,

where there is substantial scientific disagreement over proper test methods and correct interpretation of test results. CMA has stressed that the Agency should focus on developing straightforward test requirements that do not raise unresolved scientific issues.

Section 5 of TSCA is another area of much regulatory activity. Section 5 authorizes EPA to screen "new" chemicals, and "significant new uses" of existing chemicals before they are manufactured so as to identify their potential adverse effects on health or the environment. EPA's actions under Section 5 have an impact on the future development of new chemicals. Congress was well aware of the potential for stifling innovation and designed Section 5 so it would impose minimum burdens on manufacturers while still identifying and eliminating unreasonable risks as required under the Act.

Unfortunately, EPA initially proposed ambitious regulations that could have expanded premanufacture notification ("PMN") requirements well beyond the limits embodied in the statute. For example, EPA had testified in Congress that manufacturers would merely be required to complete and submit a two-page PMN form. But the PMN that EPA proposed in January of 1979 was an extremely detailed form which called for nearly 40 pages of mandatory information, and 20 more pages of optional information. CMA, and other industry commenters, urged drastic modifications in EPA's proposal in order to conform to the terms of the statute and lessen the burden on innovation. In particular, CMA developed and provided to EPA, a draft PMN form which called for much less information, but would still be adequate to facilitate EPA's review function under Section 5.

EPA responded to industry concerns by publishing reproposed regulations that were substantially more focused in scope. Nevertheless, industry still had substantial reservations about many aspects of EPA's reproposal. For example, CMA argued that EPA's reproposed PMN form was still more elaborate than necessary and called for information that EPA lacked statutory authority to require. In addition, industry expressed reservations about EPA's continued interest in prescribing supplemental reporting requirements. Finally, CMA objected strenuously to highly elaborate procedures that EPA had developed for asserting and substantiating confidentiality claims. Industry felt it was highly unnecessary to require such substantiation at the time a PMN is submitted, rather than after a request for the disclosure of PMN information had been received under the Freedom of Information Act ("FOIA") -- which is how it is done under most other EPA regulations.

While debate over the merits of its proposed regulations has continued, EPA has simultaneously received and reviewed almost 2,000 PMNs. The Agency's premanufacture review program has been shaped by the explicit requirements of the statute, supplemented by informal Agency guidance, and thus has provided an opportunity to test the industry position that Section 5 can be implemented effectively without detailed regulations.

In CMA's judgment, EPA's experience largely confirms the industry position. In general, the PMN program has been functioning smoothly and efficiently in the absence of regulations. Most PMN submitters have provided EPA with adequate information in their PMNs which reflects the fact that the chemicals have generally presented little or no risks. Submitters have almost always been responsive to voluntary requests for follow-up information. In addition, where EPA has expressed concerns about the health and environmental effects of particular chemicals, many submitters have taken informal action to allay those concerns, including voluntary suspension of the PMN review period or withdrawal of the PMN.

At the same time, however, there is emerging evidence that PMN requirements have significantly inhibited innovation in the chemical industry. Section 5 has inevitably had some impact on innovation, and EPA regulations that add to the basic statutory requirements are likely to impose burdens on the chemical industry that are unnecessary. That is why CMA recommends that the Agency, at a minimum, limit those regulations to the clear requirements of the statute. Requirements such as supplemental reporting of PMN information, mandatory consumer contacts, advance substantiation of confidentiality claims, and the "invalidation" of incomplete PMNs, will only serve to increase the costs of a process that already imposes substantial burdens. Since EPA can adequately protect human health and the environment without these requirements, they should be eliminated.

The chemical industry also encourages EPA to reexamine and rescind its Premanufacture Testing Policy. While ostensibly voluntary, this policy embodies the Agency's position that all, or nearly all, new chemicals should undergo a minimum "base set" of tests or, in OECD terms, a "Minimum Premarketing Set of Data" (MPD), which could greatly increase the costs of most PMN submittals. As Congress drafted TSCA, the decision whether to test a new chemical, and how much testing to do, resides initially with the PMN submitter and must be tailored to the particular circumstances of each new chemical. Any action by EPA which results in unnecessary testing, or prevents PMN submitters from balancing the economic and safety factors unique to their particular chemicals, will add unjustifiably to PMN costs, and needlessly inhibit innovation without a commensurate benefit.

As a footnote, this philosophy towards chemical testing is in contrast with that of the Europeans' "Sixth Amendment." The Sixth Amendment requires the development of "base set" data through mandatory testing conducted before the chemical may be marketed. The U.S. has had much more experience in implementing a regulatory scheme to evaluate the effects of chemicals, and recognizes the great costs and burdens of performing unnecessary tests. Also in the U.S., much data has already been developed that is available for evaluation of many chemicals. Further testing would be superfluous in many instances. That is why the EPA must first make a showing that a particular chemical may present an unreasonable risk or cause significant human or environmental exposure before designating it for further testing.

In addition, CMA has petitioned the EPA to establish exemptions under Section 5 of TSCA for certain chemicals. These exemptions fall into three categories -- exemptions for site-limited intermediates; low volume chemicals; and polymers. For instance, a substantial portion of the PMN filings received by EPA to date have been for intermediates and high molecular weight polymers. The risk potential of these chemicals is unusually low and they rarely require careful evaluations by the Agency. In addition, many new site-limited intermediates are typically manufactured in small volumes and are unable to absorb the costs of PMN submissions. For these reasons, CMA petitioned EPA to grant exemptions as permitted under Section 5(h)(4) of TSCA. EPA recently proposed such exemptions in August. Although some problems were raised regarding EPA's proposals during the comment period, it is likely that by 1983 the final exemption regulations will be in place.

Another important provision under Section 5 of TSCA is the "significant new use" notification requirement (called "SNUR"). A "SNUR" is designed to notify EPA when a chemical, that has undergone PMN review, presents a significant new distribution in commerce. The criteria defining a significant new use includes projected production volume, projected uses and projected exposure. A disturbing development last year was EPA's interest in using SNURs to monitor the commercial development of new chemicals that have completed the PMN review process. CMA submitted comments on the first "significant new use rule" directed at a specific chemical. Although SNURs may be appropriate in certain circumstances, CMA objected to the indiscriminate use of SNURs to follow-up on new chemicals after initial manufacture because it could subject such chemicals to a burdensome series of notification requirements throughout their life-cycles. Such recurring PMN requirements would increase the short-term and long-term costs of commercializing new chemicals, especially when more effective options are available.

I would like now to discuss a third major area of TSCA -- the recordkeeping, retention, and reporting of information under Section 8 of TSCA. The basic purpose of this information gathering mechanism of TSCA is to assist EPA in acquiring information necessary for the Agency's various regulatory activities under the Act. Section 8 is not a substantive regulatory provision. It functions as an adjunct to other provisions to provide EPA with the information relevant to the evaluation of the health or environmental effects of chemical substances.

An obvious and all-important aspect of Section 8 is that it is a mechanism to facilitate the acquisition of information that EPA needs. Accordingly, the proper test of EPA's performance under Section 8 is not the amount of information that EPA acquires or the number of companies required to report, but the Agency's success in building a data-base for accomplishing its specific risk assessment, testing and chemical control responsibilities under Sections 4, 5, 6 and 7 of TSCA. In view of this purpose, it was of great concern that EPA has repeatedly failed to define carefully, and then articulate fully, the connection between a proposal under Section 8, and a specific regulatory objective under some other provision of TSCA.

All too often, the Agency's proposals appeared to seek information for its own sake. As a result, EPA's reporting proposals often required the submission of data that are of questionable validity or limited relevance in EPA's risk assessment activities under other TSCA provisions. They also called for far more information than EPA had the capability to process and evaluate in a reasonable period of time.

An example of these problems was EPA's proposed "preliminary assessment" rule under Section 8(a). This rule would have covered 2,300 chemicals and affects a very large number of manufacturers and processors. Nevertheless, EPA's proposal failed to explain why the Agency believed it could promptly process information on this many chemicals, what precise criteria were used to select chemicals for inclusion in EPA's reporting program, and exactly how EPA intended to use the information submitted to reach conclusions concerning the risk potential of the 2,300 chemicals involved. Fortunately, in the final rule, EPA has rejected a generic approach to Section 8(a) follow-up of new chemicals and requires reporting only where it has identified specific concerns. The list of 2;300 chemicals has been reduced to 250. The rules are now structured to require reporting by manufacturers and importers, with follow-up processor reporting, which significantly reduces the burden on submitters.

Another example of excessively broad regulations is EPA's requirement of reporting allegations of "significant adverse reactions" under Section 8(c). First of all, these "allegations" are by definition unsubstantiated, and they often will be inaccurate or misleading. Moreover, broad reporting requirements might encompass many thousands of allegations throughout industry. Secondly, as written in the statute, this is a recordkeeping provision, not a reporting requirement. This proposal, again, goes beyond the statutory requirements of TSCA and has little value in achieving its objectives. We are encouraged that EPA is reconsidering some of its basic concepts under the proposal before publishing the final rule.

EPA's previous efforts to expand its authority would have resulted in the submission of data of questionable quality, and the imposition of unnecessary burdens on industry. Industry has recommended a more focused approach under which EPA could still obtain more than enough sound information to support its regulatory activities under TSCA. The reporting provisions of Section 8 are sufficiently broad to allow EPA to accomplish its objectives without departing from the basic statutory framework established by Congress.

Section 8(d) governs EPA acquisition of health and safety studies. It authorizes EPA to promulgate rules requiring any person who manufactures, processes or distributes in commerce any chemical substance or mixture, or proposes to do so, to submit lists and/or copies of health and safety studies. EPA may exclude

certain types of studies if it finds their submission
to be unnecessary to carry out the purposes of TSCA.

EPA proposed and reproposed its Section 8(d) rule
and CMA submitted extensive comments. CMA expressed
concern that the proposal applied to a broad and
indiscriminate selection of chemicals with little
attempt to explain why reporting on these chemicals is
needed. The legislative history and Sections 2 and 9
of TSCA require EPA to minimize the burden on reporters
by seeking only information necessary to achieve its
objectives under TSCA. Many of CMA's concerns were
addressed by EPA in the final Section 8(d) rule
published in September.

Although the international aspects of TSCA will be
addressed in detail by another speaker, I thought it
appropriate to mention that CMA and industry are
participating in the development of those issues raised
by the various sections of TSCA. For instance, in the
area of testing chemicals under Section 5, there are
differences between the U.S. and the European
approaches regarding when testing should be required,
which I have already mentioned.

Another area of international controversy is the
export notification requirements under Section 12(b).
EPA's regulation requires that (1) exporters notify the
EPA of their intent to export chemicals that are
subject to certain listed regulatory activities under
TSCA, (2) that EPA notify the foreign government of the
importation of such chemicals, and (3) that EPA provide
TSCA-related information to those governments. CMA
submitted a petition to EPA proposing an alternative
which provides for an annual dissemination of
information concerning all relevant regulatory
activities, listed in Section 12(b), to all foreign
governments. This comprehensive mechanism would
provide all relevant information to foreign governments
in a timely fashion. If in the interim there was a
finding that a chemical presented an imminent hazard, a
special notice could be sent. In this manner each
government could employ its own authority to regulate
chemicals imported from anywhere in the world, not just
from the United States. Furthermore, to the extent a
Section 12(b) notice might adversely affect commerce
involving a regulated chemical, all exporters operate
under similar constraints. We also suggested
procedures designed to protect confidential information
as required under Section 14 of TSCA.

The international issues raised by TSCA interface
with several arenas -- including the State Department,
Commerce Department, EPA, the Organization for Economic

Cooperation and Development (OECD), the European Economic Community (EEC) and the United Nations. CMA continues to keep up-to-date on the latest developments in these areas and is an active participant in developing reasonable and practical international policies.

As you can see, during the regulatory development of TSCA, the chemical industry has been alert to the practical impacts of the law and its ensuing regulations. Representatives of individual manufacturers, CMA and other trade associations, have participated at every opportunity to provide EPA with the factual data it requires and to show the practical implications of its proposals in order to prevent EPA from over-extending its power beyond its statutory authority. CMA has asked EPA to be more sensitive to the impact of these regulatory requirements on innovation and other economic aspects of this industry.

EPA began with overly ambitious plans which have required reexamination in light of practical and legal considerations that it first overlooked. If EPA focuses and modifies its proposals in accordance with industry recommendations, it will be able to implement TSCA in a reasonable, cost-effective, yet efficient and timely, manner.

RECEIVED December 23, 1982

Management of TSCA-Mandated Information

C. ELMER and J. R. CONDRAY

Monsanto Company, St. Louis, MO 63167

The Toxic Substances Control Act contains several
sections which require the government to collect
and manage information obtained from industry and
other sources. This paper itemizes these requirements, and reviews the status and implications of
each. Industry's experience in collecting appropriate information is outlined as are some unanticipated benefits derived from mandated information
management. Examples of one company's experience in
managing information will be detailed. Relationships
with information requirements in other countries and
their implication to U. S. corporations in responding
to TSCA regulations are discussed. EPA's requirements for broad information collection and implementation of management has resulted in only partial
compliance. Some of the future activities as well as
foreseeable problems are discussed. One response to a
TSCA mandate has been the development of a Chemical
Substances Information Network (CSIN). Although the
concept is recognized as valid it may be expanding
beyond its originally envisioned scope. Benefits as
well as potential dangers in making unevaluated information readily available to undiscriminating users
are cited with examples.

The Toxic Substances Control Act or TSCA, contains no less
than 60 authorities for developing or disseminating information. (1) These can be categorized into 15 significant classes
of information of which 8 related to requirements by industry, as
shown on Table 1.

Congress charged the Environmental Protection Agency (EPA)
to administer TSCA and, in assigning authority provided two basic
routes for implementation. In the first case, Congress spelled
out in the act what information EPA was to provide and made such

Table I

TSCA MANDATED REQUIREMENTS
FOR DEVELOPING AND REPORTING OF
INFORMATION BY INDUSTRY

Section 5(a)	Premanufacturing Notice and Significant New Uses
5(b)	Submission of Test Data as required by Section 4(a)
5(e)	Regulation pending adequate information
8(a)	General Record-keeping
8(b)	Inventory
8(c)	Records of Allegations
8(d)	Health and Safety Studies
8(e)	Notice of Substantial Risk

requirements either immediately effective or to be triggered by
some other TSCA activity.   In the second case, Congress provided
EPA with authority for information gathering.   However, in this
case, EPA must go through a formal regulatory rulemaking process
with opportunities for public comment.   In the latter case, the
submission of voluntary information to the agency is many times
preferred by both the regulated community as well as EPA.   Voluntary sharing of information is a form of mandatory information
submission.   The cooperation of the regulated community is
offered often in the hope of satisfying EPA's information needs,
without EPA issuing potentially overburdensome regulations.
Likewise, EPA benefits through closer interaction with the
information sources and the elimination of time consuming,
resource intensive rulemaking activity.

Manufacturers, processors and/or distributors of chemical
substances in the chemical industry are faced with three basic
types of TSCA-mandated information situations:

1.   Those mandated by Congress directly in the act;
2.   Those required by EPA through formal rulemaking as
     authorized by the act;
3.   Those voluntarily provided in anticipation or in lieu
     of formal rulemaking.

While these three approaches apply to many sections of TSCA,
this paper will concentrate on only three: Section 5 - Manufacturing and Processing Notices, Section 8 - Reporting and Retention of Information and Section 10 - Research Development,
Collection, Dissemination, and Utilization of Data.

Monsanto's experience illustrates both the positive and
negative impacts on one chemical company under TSCA sections 5
and 8.   For a broader perspective, refer to the Chemical Manufacturers Association publication "The First Four Years of the
Toxic Substance Control Act."(2)   The authors' views are also
presented relative to EPA's management of information under
section 10.

Pertinent Sections and Status

A brief description and status of these three sections will
help set the stage.

Section 5 deals with the notification to EPA of new substances or significant new uses of existing chemicals.   Table 2
outlines the information requirements of this section.   The
Premanufacturing Notice (PMN) requirement of the act, as required
under section 5(a)(1)(A), went into effect, as mandated by Congress, 30 days after the TSCA inventory was published, according
to Section 8(b).   Since taking effect on July 1, 1979, over 1250
PMN's have been submitted.   This activity has been one of the top
agency priorities.

Other section 5 requirements have either not been finalized
or are tied in some fashion to the PMN activity.

Table II

TSCA SECTION 5 INFORMATION REQUIREMENT STATUS

Sub-Section	Authority	Status
5(a)(1)(A)	Submit a notice, at least 90 days prior to manufacture or processing of a chemical substance not on the inventory required under Section 8 (b)	Went into effect on July 1, 1979 (30 days, after publication of inventory) as mandated by Congress. To date over 1250 notices have been submitted. EPA published guidance in the form of interim policy (44FR63006) and several proposed rules, the most recent being (44FR59764).
5(a)(1)(B)	Submit a notice at least 90 days before an "existing" chemical substance can be manufactured or processed for a use that EPA has determined by rules, is a "significant new use"	Not finalized. Proposed rule issued for N-methane-sulfonyl-p-toluenesulfonamide which would require a notice if the volume exceeds 1000 pounds, (45FR78970).
5(b)(1)	Submit test data with notice as required under Section 5(a)(1)(A) if covered by a Section 4 test rule	Went into effect July 1, 1979 when 5(a)(1)(A) activated. No submissions to date.

Table II (cont.)

Sub Section	Authority	Status
5(b)(2)	If substance in question is on Risk List under section 5(b)(4) then, submit data with notice as required under section 5(b)(1)(A) which shows substance with respect to notice will not present and unreasonable risk	Went into effect July 1, 1979 when 5(a)(1)(A) activated. No submissions to date (see 5(b)(4)).
5(b)(4)	The Administrator may rule, compile and keep a current list (Risk List) of substances that may be present and unreasonable risk of injury to health or the environment	No action.
5(d)	Defines the content of a notice under section 5(a) to be described in Section 8(a)(2) A-D and F-G	Went into effect July 1, 1979 when 5(9)(1)(A) activated.
5(e)	Require a submitter of a notice under section 5(a) to submit additional information if EPA has insufficient information and have reason to believe the substance may present an unreasonable risk of injury to health or environment	EPA has issued several 5(e) orders. None has been challenged, and all notifiers to date have withdrawn the notices rather than submit additional information.

(Cont. on next page)

Table II (cont.)

Sub Section	Authority	Status
5(h)(1)	The administrator may upon application exempt any person from the notification requirements of section 5(a) or (b) if the substance in question is limited to test marketing	Numerous test marketing applications have been submitted and approved.
5(h)(4)	The administrator may upon application and by rule exempt the manufacturer from all or part of the section 5 requirements	Applications for exemption have been filed by individuals and several industry associations. Nothing finalized to date.

Section 8 is the primary information gathering authority
under TSCA. Requirements and implementations status is shown on
Table 3. Section 8(a) grants broad authority to the Administrator
of EPA and permits EPA to obtain reports as they may "reasonably
require." EPA's only burden is to request the information through
a formal rulemaking procedure. To date, EPA has not finalized any
rulemaking under section 8(a) except reporting requirements for
two specific chemicals. (TRIS and PBBs). The agency has actively
pursued section 8 implementation. However, it has finalized rules
for only the inventory requirements (section 8(b)) and has
organized to handle the management of substantiated risk notices
mandated by Congress under section 8(e).

This paper does not address EPA plans for further implementation of sections 5 and 8 of TSCA. The agency is active in both
areas. For information on future plans refer to the most recent
EPA publication of its events calendar[3] and EPA Office of Toxic
Substances report Priorities for OTS Operation[4].

Section 10 is a Congressional mandate to EPA for management
of the massive amounts of information that flow into the agency
under TSCA authority.

Under Section 10(b)(1), responsibility is assigned to an
interagency committee to establish, within EPA, an effective
system for collection, dissemination to other federal departments
and agencies, and use of data submitted under TSCA.

Paragraph (2) of the same section calls for establishing an
effective system for retrieval of toxicological and other scientific data which could be useful to the administrator in carrying
out the purposes of TSCA.

Under section 10 authority, EPA has a Chemical Substances
Information Network (CSIN) prototype under test by industry,
academia, unions and state government offices. The system is
intended to permit easier access to existing data resources.

Monsanto's Experience

Monsanto's experience with management of information mandates under sections 5 and 8 fall into the three categories
as outlined above:
(1) Compliance with TSCA (Congress) mandates;
(2) Compliance with EPA TSCA rulemaking;
(3) Voluntary submission and actions.

Monsanto, like many chemical firms, had a product safety
program in place long before passage of TSCA. The Monsanto
program from its inception embraced the spirit and in many cases
the letter of what later appeared as the law. In some situations
TSCA has improved the focus of Monsanto's information management
and has been beneficial. In other cases we have provided constructive criticism of EPA proposed rules which required information submission and management with no specific objective.

TABLE III

TSCA Section 8 Information Requirement Status

Sub-Section	Authority	Status
8(a)	Retain and submit to EPA reports as the Administrator may require	Not finalized. Proposed rule issued which would require submission of general exposure information for 2300 chemicals (45FR13646).
	Submit notice of manufacture or import of Tris (2,3-dibromo-propyl) phosphate and Polybrominated Biphenyls	Final rule issued 44FR33525. Proposed rule for record keeping reporting of Asbestos issued 44FR8200. Not finalized.
8(b)	Compile, keep current and publish an inventory of each chemical manufactured or processed in the United States	Final rule issued 42FR64572. Inventory compiled and published. Inventory kept current by EPA and updates periodically published, the most recent issue June 1980.
8(c)	Retain and submit to EPA records if significant adverse reactions to health or the environment, alleged to have been caused by a substance or mixture	Not finalized. Proposed rule issued which would require automatic reporting after 3 allegations received (45FR47008).

TABLE III (Cont.)

Sub-Section	Authority	Status
8(d)	Submit lists of health and safety studies conducted or initiatied or known to, or reasonably ascertainable and when required submit actual copies of studies	Not finalized. Final rule issued (43FR30984) but revoked (44FR6099) Rule reproposed which listed 67 chemicals or categories of substances which would require reporting (44FR77470).
8(e)	Submit notice to EPA of information which reasonably supports the conclusion that a substance or mixture presents a substantial risk of injury to health or the environment.	Immediately effective on passage of TSCA October 11, 1976. EPA published guidance for compliance (42FR45362) and (43FR11110). Over 400 notices filed to date.

The first specific information activity initiated under TSCA
was the development of a policy and procedure for handling section 8(e) notices. It went into effect immediately after passage
of the act since this requirement was a congressional mandate. A
corporate system was established to permit individuals in the
organization to discharge their personal responsibility by reporting up through a communications channel. The system is still
in place and functioning. Since trivial concerns that enter the
system must be processed, it has placed a burden that did not
exist under our previous product safety program. The process
has, however, improved communication flow.

The 8(b) inventory accumulation was the next major activity.
For a decentralized company like Monsanto or, for that matter,
most major chemical companies, the experience of centralized
information gathering was a new experience. However, we believe
that the experience not only was novel, but proved to be beneficial from several points of view. First of all, it enabled us
to evolve a network of expertise. Second it gave us a central
data-base on which to build other information important from a
corporate point of view, and permit a one-time expense for developing a system. Third, it revealed that we needed to improve our
data files in some areas. And, fourth, it gave our central staff
departments some surprises as to substance locations. We used the
Chemical Abstract Service Registry Profile capabilities to gather
all the known synonyms and added our internal numeric and common
identifiers to access the file via dozens of possible names or
numbers.

The inventory of products, isolated intermediates, imports
and useful byproducts was initially collected in a complex data-
base with room for additional substances and attributes. For
TSCA submission, computer tapes were then easily produced with
appropriate plant location grouping and CAS Registry number and
inventory number identification. This not only saved considerable clerical effort but assured accuracy in transcription
and form preparation.

For Monsanto, the core inventory naturally led to an expanded corporate inventory which was placed into this new data-
base. It now includes raw materials, supplies and a considerable
list of other types of substances and physical agents encountered
in the workplace.

The original inventory file has grown into a wide ranging
data-base with utility for emergency response, Material Safety
Data Sheet information and is the cornerstone for Monsanto's
Occupational Exposure and Medical Systems. Therefore, the
original centralized effort has borne considerable fruit in
addition to and independent of the TSCA inventory.

Monsanto's experience with information-gathering and reporting under section 8(d) - Health and Safety Studies - has not been
as rewarding as the inventory activity. On July 8, 1978 EPA
issued a Final Rule requiring submission of lists and copies of

Health and Safety Studies for the chemicals and chemical categories included on the first two Interagency Testing Committee
Priority lists. Monsanto manufactured, processed or distributed
several materials on the list including some that fell into the
categories given. Unfortunately, EPA did not define the scope of
categories listed. This led to a great deal of confusion as to
the coverage of the rule. Monsanto's submission for one category,
the alkyl phthalates, numbered over sixty reports. A great deal
of clerical work was necessary to organize, reproduce and to
submit the one-foot-high stack of reports. We later found out,
much to our concern, that this information had not been made
available to all groups in EPA responsible for alkyl phthalates.
Monsanto benefited from EPA's rulemaking as a result of the
mandated organization activity. However, the primary objective
of making the information available to EPA for use was not
accomplished.

On January 31, 1979, EPA revoked the 8(d) rule and has
since reproposed a new rule. Monsanto's comments on the
reproposal address the concerns experienced with the first
rule and other suggested modifications.

Monsanto's experience in implementing Section 5(a)(1) - PMN
requirements - again highlighted the need for centralization.
While existing product safety programs for new products were in
place, a need surfaced to assure corporate uniformity, to prevent possible duplication and to coordinate internal information
management and EPA submissions. A central staff department now
provides a TSCA management function to assist in submissions and
coordination of all documentation. This has worked well for the
PMN's submitted to date. We have not waited for regulatory
mandates as to the type and amount of toxicology testing to
develop or submit, but have followed our internal testing policies prior to commercialization of a product.

All Monsanto PMN's submitted to date have voluntarily included information beyond that mandated by the act. This has,
in some cases, included: Material Safety Data Sheets, Label
information, details regarding the industrial hygiene programs
in the proposed manufacturing site(s), risk assessment information and other related information. While this information
is not mandated, we believe in most situations it is to our
benefit to assist the agency in this way.

EPA's Information Management

Information management by EPA, as mandated under TSCA has
centered on the development of CSIN (Chemical Substances Information Network). In response to Section 10(b) this information
management system has evolved to retrieve toxicological and
other scientific data which could be useful to the Administrator
in carrying out the act. After several years and about an equal
number of millions of dollars, currently a prototype networking

system is under evaluation by up to 50 organizations. A wide ranging user group - in terms of both interest and capability-is providing a first pass to establish utility of the system.

Simply put, the CSIN concept appears to be pure information management. It does not create or store data or references. It points and selects - with a broader choice of data-bases and via easier techniques than heretofore possible. An analogy of a library might be useful. Visualize a central catalogue for a vast interdisciplinary library with book and journal holdings which contains all references to a particular subject regardless of the particular source document. Access to the system is through a terminal which responds to relatively easy in-structions. Current data-bases would be compared to many independent libraries, in different locations, with indices which have different methods for searching. CSIN enables the user to search all of these via a centralized, unified, methodology.

CSIN administration and planning have been performed by about 25 federal agencies, including EPA, DHHS, HTP and CEQ. EPA has played the central role in implementation of plans.

Two concerns arise which CSIN administration has not addressed so far. First, the emphasis has been to provide access only through well-established existing sources. No apparent effort has been made to consider the need for drawing together the multiplicity of information submitted to the EPA Office of Pesticides and Toxic Substances in an easily accessible form for agency use. With the exception of information contained in the Chemicals In Commerce Information System (CICIS), developed pri-marily to accommodate TSCA Inventory and other related infor-mation, regulatory personnel often are not aware what is already available and request repetitive submissions from industry. Furthermore, conclusions are drawn only from publicly available data—bases. Data already available within the agency are not readily accessible for its own personnel. Apparently, this situation is a result of non-responsiveness, so far, to TSCA Section 10's mandate.

The second concern or problem is more difficult to deal with and could have more far-reaching implications. It involves the lack of discrimination in the choice of information files made available through CSIN as well as the lack of concern about the reliability of the information received. The greater the tech-nical expertise of users, the less concern for these questions since this group will seek the original source and discriminate between "good" and "unreliable" data. CSIN, however, attempts to make exhaustive searches available through simplified method-ologies with the implication that the unsophisticated or non-technical user can readily obtain "all" available data.

Our experience has shown that computerized searching for information results in a false sense of credibility for what is retrieved. Additional confidence, particularly on the part of

non-technically trained searchers is gained since this is a
government created system. We know this doesn't necessarily
make sense or sound rational. However, we are dealing with
people and subjects which are not always rational. A need to
substantiate or justify an emotionally based opinion via
so-called "hard" data, regardless where it comes from or how
reliable it is, can result in wide-spread and far-reaching
implications.

As an example, the NIOSH-RTECS (Registry of Toxic Effects of
Chemical Substances) is a reasonable basis for beginning a search
regarding the toxicity of a substance. The printed version
contains some 47,000 materials and its contents are thoroughly
documented in its introduction and file description. The most
commonly reported effect is the LDLo or lowest lethal dose found
in a species or the LD50 which is the lethal dose for 50 percent
of the group under test. ONLY THE LOWEST DOSES ARE PRESENTED IN
RTECS, REGARDLESS OF TEST OR LABORATORY RELIABILITY. To the
unskilled, these selection criteria can easily be missed. and
critical judgements made on the basis of one highly variable test
method.

The file lists all chemicals on which toxicity data have
been found. In other words, the title is not: Registry of Toxic
Substances. Yet, some believe that all substances in the Registry are necessarily toxic under any conditions of use. Some
state regulations have even been drafted based on this belief.

Computerized data-base versions of RTECS give the capability
of extracting substance lists by "Classification Codes." For
example, one might ask the system to search for all compounds
with classification Code of TUMORIGEN. How many users will have
read the user's guide carefully enough to know that this means
only that these compounds may have been reviewed by IARC or NTP
but NOT that they have been indicted as tumorigenic?

Now, what can EPA or anyone else do about this elusive but
real problem? A start has already been made via a data quality
workshop which was initiated by the CMA (Chemical Manufacturers
Association) and co-sponsored by EPA, NBS (National Bureau of
Standards) and NAS (National Academy of Sciences). This resulted
in a group of about 40 experienced participants from government,
industry and academia reviewing criteria for data quality in four
areas of information relating to properties, health and environmental effects. From this beginning, we eventually hope to see
the contents of data-bases or data files identified as to the
level of reliability of extracted information. The user will
then at least have the ability to judge the value of the
information he received.

Conclusion

The vast majority of major,chemically-oriented companies has maintained on-going, socially responsible attitudes towards the public and the environment in relation to their products long before TSCA came into being. This, despite the hue and cry of the media and certain public factions who continue to feel that industry's concern for health and the environment must be legally mandated. Toxicity and environmental evaluations go back at least 20 years and, in some cases, forty years, using best available science and techniques available at that time. Material Safety Data Sheets, describing hazards and precautionary measures, have been produced for an equal period. Reviews of new products prior to commercialization have been mandatory in our company well before PMN rules appeared.

Laws and regulations by their very nature place constraints and burdens on the regulated community. TSCA is no exception. Monsanto's experience with sections 5,8 and 10 of TSCA has, however, indicated that in addition to the burdens imposed by the mandate, there are also tangible benefits. Such benefits help offset some of the burdens and satisfy the intent of Congress as indicated in Section 2(c) of TSCA. "It is the intent of Congress that the Administrator shall carry out this Act in a reasonable and prudent manner, and that the Administrator shall consider the environmental, economic and social impact of any action the Administrator takes or proposes to take under this Act."

The challenge facing EPA, industry and the public interest groups is to design and implement information management rules and systems that satisfy both the needs mandated by TSCA and the congressional intent on economic impact. To quote Monsanto's Vice Chairman, Dr. Louis Fernandez, from a recent speech to the National Association of Manufacturers, "it's clear that industry is committed to and capable of achieving our nation's environmental goals. We hope the regulatory agencies have come to recognize this and that we can work together to shape effective and reasonable regulations in the future".

LITERATURE CITED

1. Chemical Reporting and Recordkeeping Authorities under 15 Environmental and Consumer Acts, (EPA Report 560/3-78-001) 1978
2. Frost, Edmund B., Cox, Geraldine V., Hutt, Peter Barton The First Four Years of the Toxic Substances Control Act (A Review of the Environmental Protection Agency's Implementing TSCA,) Chemical Manufacturers Association 1981
3. 46 FR53995 Environmental Protection Agency Regulatory Agenda, Toxic Substances Control Act, October 30, 1981
4. EPA Office of Toxic Substances, Priorities for OTS Operation Office of Toxic Substances, U.S. EPA, January 1982

RECEIVED September 1, 1982

9

Impact on Corporate Structure and Procedures

DAN R. HARLOW

Diamond Shamrock Corp., Washington, DC 20006

Corporations have responded to the Toxic
Substances Control Act (TSCA) by traditional means
such as added personnel and facilities as well as
by new and innovative initiatives such as new
management functions, new expertise in toxicology,
and new computerized information systems. The
size of a corporation as measured by annual sales
is the most important determiner of how an
organization responds to TSCA's new demands; large
corporations generally add new personnel and
facilities while smaller corporations tend to add
TSCA requirements to existing jobs. None of these
responses can be ascribed to TSCA alone since
corporations are responding to a myriad of
environmental/health laws with similar demands.

The Toxic Substances Control Act (TSCA) was signed into law
in late 1976. It was a bitterly-debated piece of legislation;
it arose at the height of the combined environmental concern and
"cancerphobia" fears distinctive of that time. In order to
understand how corporations have responded to TSCA, it is
important to recall the situation at that time in the knowledge
of carcinogenesis, in the environmental movement, in the
political scene and, finally, in the corporations themselves.

The author feels it his responsibility at this point to
apprise the reader of the fact that there is little "hard" data
on corporate responses to TSCA, especially at the management
level. Hence, this paper has taken the role of an overview and
qualitative look at corporate responses to TSCA rather than a
quantitative document based on estimated numbers of personnel
added and other costs of complying with TSCA. Such "hard" data
approaches on various specific aspects of TSCA such as
pre-manufacturing notification and inventory have been attempted
with some successes and some failures. At this overview level,

0097-6156/83/0213-0121$06.00/0
© 1983 American Chemical Society

the ability to ascribe particular responses entirely to TSCA or apportion that part of the response due to TSCA alone is difficult if not impossible. To the extent that this results in an anecdotal and personal approach to the impacts of TSCA with no cited references, the author apologizes for this shortcoming.

The History

The Toxic Substances Control Act was a combined product of the environmental movement and the cancer fears distinctive of that era. Much of the health emphasis of the Act fell on cancer. Huge resources were allocated to "conquer cancer", especially through federally-funded programs such as the National Institutes of Health. (By the mid-1970's the National Cancer Institute, one of eleven institutes, had risen to claim more than half of the entire NIH budget and have the only politically-appointed institute director.) President Nixon announced that if we could go to the moon, we could certainly conquer cancer, and do it within seven years! The prevalent models for cancer etiology in the early 1970's were the virus model and the chemotherapy approach, both of which had lost the promise of providing the overall answer to carcinogenesis. Hence, as the chemical modification of DNA model gathered momentum, it became attractive to the National Cancer Institute as its new hope for the answer. After it was stated in the early 1970's that 60%-90% of all cancers are environmentally induced, the stage was set for chemicals and environment as the issues of focus. We still don't know the etiology of cancer--we do know that some specific stimuli such as some viruses, certain kinds of radiation and specific chemical stimuli can cause a variety of different malignant diseases under certain conditions perhaps by activating to oncogenes found in normal cells.

As early as the 1930's, acute toxicology testing was recognized as a necessity for companies dealing with substances which might present acute health hazards. Indeed, industry became a leader in building the field of acute toxicology. With cancer, however, science itself was reticent in accepting chronic testing. The Millers of the University of Wisconsin, pioneers in chronic testing, recently stated that in the 1940's when they began chronic testing in their laboratories that they were looked upon with suspicion and some disdain even by toxicologists (who were then involved only in acute testing) to say nothing of the lack of regard expressed by other scientists. The often-repeated statement that most cancers are due to environmental stimuli (environmental was originally intended and recently emphasized to include such factors as diet, smoking, lifestyle, stress, etc.) capped the cancer fear and sent us headlong into a massive "search and destroy" mission for environmental carcinogens. Particularly suspected among environmental stimuli were manufactured chemicals. This is

largely true today, although recently initiated cancer studies
are focusing on diet and lifestyle factors which will likely
change the emphasis of "search and destroy" missions in the
future.

In the change of emphasis from acute to chronic effects
many scientists were unprepared to deal with the new concerns.
When the full force of the cancer fear and targeting of
manufactured chemicals hit, the science of chronic testing and
much of the chemical industry were not well prepared. Public
concern was expressed through a variety of laws including the
Toxic Substances Control Act. The industry responded to the
challenge in a variety of ways according to company unique
characteristics. Those companies which had strong acute testing
programs were better prepared to deal with demands for chronic
testing than those which had less active acute programs. Some
companies were aware of the growing need for chronic testing and
were on the leading edge of its development while others were
less aware. Those companies which had dealt with occupational
hazards to highly toxic, acute or even fire and safety hazards
were more likely to be monitoring workplace concentrations of
various substances and had industrial hygiene programs in place;
those that did not deal with such hazards did not.

The science of chronic testing was not well-formed in the
years just preceding the implementation of TSCA and even today
is undergoing modifications and re-evaluation. For instance, at
the time of TSCA implementation, the National Cancer Institute
bioassay was considered the "state of the art" for cancer
testing, although the NCI itself considered it only a screening
mechanism. This bioassay procedure, based on MTD (maximum
tolerated dose) lifetime exposure to rodents, has undergone
modifications and criticisms to the extent that many of the
bioassays done in the early 1970's are no longer considered
reliable. In addition, new modes of testing have arisen since
TSCA's debut, making the job of chronic health testing a "moving
target." These changes are not decreasing with time; in fact,
there have been demands to develop new, faster and less
expensive testing methods for carcinogens. Similar changes and
pressures have been seen in environmental testing which, in
general, is even less well-defined than carcinogen testing.

At the same time that this increased awareness to
environmental carcinogens arose, the environmental movement was
in full bloom. Since its established political goals were quite
consistent with environmental carcinogen fears, the two were
quite complementary. The history and development of the
environmental movement are a fascinating and complex story in
itself; however, it is not an appropriate subject for this
paper. Let it suffice to say that it was the combination of the
cancer fears and the environmental movement that gave the
political strength to pass TSCA into law.

Thus, at the time of implementation of TSCA, companies exhibited a wide variety of sophistication and understanding of what TSCA would demand from them, and the science of carcinogen and environmental testing were in the development stage. What kinds of responses have been generated from these pressures? I will try in the following pages to characterize the responses of some companies. One must remember that we cannot at this time write the requiem for how companies have responded to TSCA since corporations are still responding to the on-going implementation of TSCA.

The Response

The institutional resources which a business enterprise can devote to the pressures and demands of TSCA generally fall within five large categories; a variety of other internal activities, and external activities.

- Personnel - new staff support functions added
- Testing Facilities - new testing facilities added
- Recordkeeping and Surveillance Programs - computerized or non-computerized, new computer software programs or manual recordkeeping procedures
- Research and Development - new product development
- Compliance Burdens - impacts of "bans" under TSCA
- Other Internal Activities
- External Activities

These additional personnel, facilities, programs and other expenses, of course, exist in the corporate structure as added costs of doing business and are accounted for in the cost to the consumer of the business enterprise's products and services.

Personnel. More than any other area, we are often asked: "How many people has your corporation added due to TSCA?" I don't know of anyone who has a concise answer to this question. Complicating the situation is the fact that the 1960's and 1970's saw a number of environmental and health laws go into effect: the Clean Air Act, Clean Water Act, Occupational Safety and Health Act, Safe Drinking Water Act, Federal Water Pollution Control Act, TSCA, Federal Food, Drug and Cosmetic Act, Hazardous Materials Transportation Act, Federal Insecticide, Fungicide and Rodenticide Act, Resource Conservation and Recovery Act, and Comprehensive Environmental Response, Compensation and Liability Act, to mention the major ones. This mixture of acts, with some similarities of purpose, developing within a time span of 10-15 years, has made a variety of similar demands. It is not easy at this point to attribute the addition of staff support personnel to an individual law such as TSCA. The same observation is applicable to all corporate resources which have felt the effects of TSCA; however, in order to prevent repetition it will not be restated in the following pages.

TSCA has resulted in increased staff support in five areas:

(1) Legal - interpretation of the law

(2) Technical - understanding of technical aspects of the law

(3) Engineering - knowledge of control procedures for manufacturing processes, etc.

(4) Toxicology/Testing - toxicology/testing advisors to management and for testing requirements

(5) Management - new management functions for TSCA

(1) Legal. It is common for large companies to have the TSCA responsibility assigned to a specific staff attorney or attorneys. This may be a full-time job in the largest companies or may be a part-time function. In smaller companies it is certain to be a part-time function among the many jobs of the few attorneys. In even smaller companies, who may not have a legal staff at all, there may be an outside counsel involved, if there is any legal function at all. These small firms may depend on trade association legal support and/or a non-attorney in the company with a basic knowledge of the law.

Rather than discuss a group of specific corporations and their responses to TSCA (which vary greatly depending on their individual structures, product mixes, sizes, number of employees, etc.), a range of corporations from small to large based on annual chemical sales was selected as examples. This range, as you will see, has a significant impact on how they have responded to TSCA. (Very large=$5 billion annual chemical sales; large=$1.5 billion; medium=$650 million; small=$20-50 million.) It has been suggested that an additional category--"extremely small=$1-10 million"--be added. It was found that these are often one or two manager operations and, thus, all of the TSCA burden falls on the one or two managers at the expense of their other vital duties. It should therefore be noted that this group of companies was deleted from the following comparison charts only because the entry would be repetitive; that is, the one or two managers must pick up the added burdens. It can be persuasively argued that these "extremely small companies" have indeed experienced the greatest impact from TSCA and find their businesses imperiled by its added burdens. The reader is asked to mentally add the "extremely small" company and its burden on its few managers to each comparison chart.

Firm 1 - Very large. Part-time (half-time) attorney at corporate, five half-time attorneys at operating companies.

Firm 2 - Large. One attorney assigned responsibility; at times full-time, now part-time. General counsel office support as needed.

Firm 3 - Medium. Part-time basis for two in-house
attorneys.
Firm 4 - Small. No in-house attorney, Research Director
handles TSCA matters.

(2) Technical. Additional technical support in companies
other than testing personnel will arise from the need for
additional recordkeeping, such as the annual reports and TSCA
inventory (§§8(a) and (b)), records of significant adverse
reactions to health or environment (§8(c)), records of
applicable health and safety studies (§8(d)), and notices of
substantial risk to health or environment (§8(e)). In larger
companies, chemical technical personnel may work with computer
personnel to put the information into computerized
recordkeeping. In smaller companies, there may or may not be
computerized services; in the small companies there will not be
computerized recordkeeping but will be some form of manual
records generally kept by technical personnel.
    In addition to direct burdens of recordkeeping which fall
on technical and technical management personnel, another burden
of major significance is the interpretation and determination of
potential impacts of TSCA on a corporation. Due to the highly
complex, technical nature of TSCA, the major job of
interpretation and impact analysis falls upon technical, not
legal, experts. This burden is essentially universal, impacting
all companies regardless of size and other characteristics. The
primary significance here is that either these TSCA duties
detract from the other technical duties (in smaller companies
even research and development activities) or result in new
technical expertise being added to existing staff.
    Firm 1 - Very Large. One full-time technical person at
            corporate level, five part-time at operating
            companies. About five other staff personnel in
            various activities.
    Firm 2 - Large. One technical persons in corporate
            headquarters; about 15 people part-time in
            divisions of the company.
    Firm 3 - Medium. One technical person in headquarters,
            four to five people in other parts of the
            company.
    Firm 4 - Small. Research Director maintains records.

(3) Engineering. Environmental engineering staffs have
had to respond to the many environmental laws as detailed
earlier. TSCA has had a minimal additional impact here since
the primary focus of TSCA is information and testing. Engineers
often participate in reporting, such as §8(e), in larger
companies with engineering staffs. In the smaller companies
without such staffs, engineering consultants provide the service
or the technical staff performs the function.

Firm 1 - Very Large. In-house engineering staff provides service on an as-needed basis.

Firm 2 - Large. In-house engineering staff provides service on an as-needed basis.

Firm 3 - Medium. In-house engineering staff provides services on an as-needed basis.

Firm 4 - Small. At this point, engineering is not needed.

(4) Toxicology/Testing. Perhaps the greatest variety in response to TSCA is in the testing area. The variety generally revolves around the question of in-house testing facilities versus external testing facilities. Among the largest companies with massive testing facilities, a large amount of in-house testing can be expected, but even in these cases the demand for testing in general (not just TSCA) has overloaded the system. Hence, much testing is done outside the companies' facilities and added personnel to monitor and assure the quality of outside testing can be attributed to other demands as well as TSCA. The general demand for ability to analyze toxicological data has resulted in the employment of toxicologists in the large companies who, in addition to their specific duties in testing, also advise management and operating units on various aspects of TSCA and other laws. For instance, they will likely be involved in substantial risk determinations under §8(e). Smaller companies do not employ toxicologists or other testing personnel but contract outside testing and toxicology consultants.

Firm 1 - Very Large. Toxicology staff of 25 provides advice on all aspects of toxicology including TSCA.

Firm 2 - Large. Two toxicologists full-time and one part-time at corporate level for TSCA.

Firm 3 - Medium. Toxicology testing and expertise outside.

Firm 4 - Small. Toxicology testing and expertise outside.

(5) New Management Functions. The infra-structure of support described above to provide TSCA-required services is all within the staff functions of the specific companies. As a result of the uniqueness of corporations, staff personnel exist in a wide variety of locations: at one extreme, they may be a highly centralized corporate staff department with line authority to operating units; on the other extreme, they may be detailed out to the operating units themselves reporting to the operating unit managers. There is no pure and simple company example for each of these two extremes; companies are mixtures of the two with a preponderance in the large companies of a corporate staff reporting to the top corporate managers with no line authority to operating units. In smaller companies, the reporting is directly to the corporate officers themselves.

In large companies there has generally been the emergence of a new corporate staff department of environment and health with a corporate level vice president (either a staff vice president or vice president/corporate officer). This job is generally a relatively new one and is often the focus point for the staff infra-structure not only for TSCA but for all environmental, safety, industrial hygiene and medical affairs. Again, how corporations structure this reporting varies with individual companies. In some larger companies, the TSCA-related requirements are sufficient to justify the existence of a "TSCA Co-ordinator" or "Director of TSCA Regulation," etc. who reports to the vice president. This person's responsibility is to focus TSCA matters corporate-wide; advise the operating units on TSCA matters which affect them; report the results and progress in TSCA matters to the vice president who in turn reports to the top corporate officers. In smaller companies, the management of TSCA is added to the other business functions of the few corporate managers.

Firm 1 - Very Large. Senior vice president (member of board) heads all health and environment staff. Full-time TSCA coordinator at corporate level, part-time Washington office support.

Firm 2 - Large. Staff vice president manages health, environmental and safety, including TSCA.

Firm 3 - Medium. Vice president, environmental health and safety.

Firm 4 - Small. Director of Government Affairs function added to existing Research Director.

Testing Facilities. As mentioned, there is wide variation among companies as to in-house testing capacities. Some companies, even large ones, rely entirely on contracted outside testing for a variety of reasons; e.g., overhead management costs, acceptance of results by the government and the public, etc. Smaller companies have used outside contractors for testing where required. There is no doubt that TSCA is requiring additional testing--the only question is where it is being done.

Firm 1 - Very Large. Very large in-house testing facilities, about 50% TSCA-related in-house, about 50% contracted out.

Firm 2 - Large. Limited facilities for in-house toxicology testing; all TSCA-related testing contracted out.

Firm 3 - Medium. In-house testing facilities for chemical and physical properties, not toxicology.

Firm 4 - Small. In-house testing facilities for chemical and physical properties, not toxicology.

Recordkeeping and Surveillance Programs. As with testing, the degree of sophistication varies greatly among companies. A small company with a few products and only a few employees cannot justify a sophisticated computer soft-ware program while a large company would be foolish to try to keep such records manually.

In large companies with computerized capability, if operating units vary considerably in the kinds of materials used, TSCA information may be kept on a unit basis and not at a large central corporate facility. However, there are several systems designed for central corporate use which bring together the entire breadth of the corporate activities for TSCA, industrial hygiene, environmental monitoring, etc. For example, Diamond Shamrock Corporation has developed a patented soft-ware system for sale trademarked "COHESS®" (Computerized Occupational Health/Environmental Surveillance System). COHESS brings together on a corporate-wide basis three separate modules--people, places and things--for a centralized recordkeeping function. The things module is a list of all chemical substances handled in the company; places is a grid network system to allow identification of any point in any plant site or facility; and finally the largest module is people. The people module collects health incidents for each employee (clinic visits, health insurance claims, workmen's compensation, accident/safety incidents, medical absence and death certificates) and health evaluations for each employee (pre-employment physical exams, annual physical exams, periodic exams, special evaluations and personal sampling in the workplace). The three modules are interconnected. For example, the people file contains the grid numbers (as used in the places module) of the specific workplace for each employee. In the things module, which can be the TSCA inventory, each chemical substance is listed with the grid numbers from the places module where that particular chemical is used. Also included in the things file for each chemical are amounts and modes of environmental exposure and amount of human exposure. As one can see, there are many ways to use such a centralized computer file. If, for instance, a particular employee shows symptoms, the COHESS file can tell what his grid location is and from that determine the chemicals he or she is exposed to. In the reverse, if the company suddenly discovers or is told that a particular chemical substance is a health hazard, the COHESS system can display the points in all the facilities where that substance is used and identify all employees working in those areas. These relatively simple maneuvers can be carried out in minimum time with a system such as COHESS. More complex retrospective epidemiological studies and long-term environmental studies can also be performed using this system since it accumulates information throughout the time it is in use.

Other companies have developed various types of computerized systems to achieve similar objectives. It should be pointed out here that the COHESS project was started in Diamond Shamrock in 1973, three years before TSCA became law. Thus, it is difficult to say that it was developed in response to TSCA. It does, however, respond to many of the recordkeeping requirements in TSCA.

Companies without computer capability and smaller companies generally must stay with manual recordkeeping, at least for the present time. Several years ago, some consultants offered to "pool" small companies until size was sufficient to make computerization possible. Potential customers rejected this option out of fear of losing trade secret information on mixtures and processes.

> Firm 1 - Very Large. Very complex centralized computerized system including medical records, industrial hygiene, environmental effects, toxicology and materials (TSCA inventory).

> Firm 2 - Large. Centralized corporate-wide computerized system for workers, workplace and employees.

> Firm 3 - Medium. Computerized records kept on substances (TSCA inventory) and on industrial hygiene.

> Firm 4 - Small. Manual records kept.

Research and Development. Among the most discussed impacts of TSCA on companies is the impact on innovation. With PMN review, companies have been required to factor in environmental and health concerns very early in the product development cycle; that is, at the research and development stage. A potential product can be dropped due to negative environmental or health information at various stages, starting at the preliminary review of existing knowledge of the potential product. As the product moves further and further along the process, all existing knowledge about the material must be gathered and evaluated until the point of decision on commercialization, when there may be a decision to test or not test, depending on what is known about the substance.

The research and development departments draw on all other personnel and testing capabilities, as discussed above, in achieving this review. Research and development personnel are made aware of the impacts of TSCA in their routine functions. As it is now a part of the R&D cycle, it would be very difficult to assign a specific burden on personnel and facilities since it reaches into various parts of existing companies as described above. In general it is believed that TSCA has resulted in a slow down in developing new chemicals.

Compliance Burdens. Under §6 of TSCA, the banning of particular chemical substances has had a significant impact on companies who manufacture those substances. For example, the banning of polychlorinated biphenyls (PCBs) has had a sizeable impact on the manufacturers of these products. A similar situation occurred with banning of chlorofluorocarbons in certain less critical uses. The "downstream effects", that is, the economic effects on users of these substances has also been substantial since more expensive and/or less effective substitutes must be used. The effects of such banning in the chemical specialties markets has been striking with existing companies being severly damaged and new companies being created. In some cases such as the PCB issue, the economic impacts went far beyond the chemical industry and its user industries to the utility industry where the final economic impact will be borne by the residential electric ratepayer and consumers of goods produced by electricity.

As §6 actions continue, such "cascade effects" as the PCB situation will be seen; the first stages impacted will be the manufacturer whose product line will be modified. The next level of effect will be the user industries who are the primary consumers; the final economic impact will fall on the consumer of the product or service.

Other Internal Activities. Under §8(e) of TSCA a company must report "substantial risks to health or environment" to EPA. The guidance published for §8(e) committees exist in most large companies and there is a regular receipt by EPA of §8(e) notices. As other sections of the Act are being implemented, companies will institute internal review committees to handle these particular requirements.

External Activities. The chemical industry responds to TSCA through a variety of activities external to the individual corporation or company. Such organizations did not arise as a result directly of TSCA but do respond to some of TSCA's requirements. For example, the CIIT (Chemical Industry Institute of Toxicology) carries on testing of various chemical substances for the chemical industry, thus preventing duplicative testing by each manufacturer or user. In addition, its separate identity from the chemical industry increases the acceptability of its test results.

The American Industrial Health Council (AIHC) was created in 1977 to address generic chronic health issues and has interests in chronic health testing such as may be used in TSCA §4 in a general sense. In addition to these two specific organizations, the wide range of trade associations serving particular industry segments have involved themselves with various aspects of TSCA.

The impact on corporations of these external activities is the demand upon corporations for employee time to work with the organizations. This so called "sweat equity" or contribution of time of key corporation employees including officers and CEOs

(Chief Executive Officers) imposes a significant burden on corporations.

Summary

The chemical industry has responded to the demand presented by TSCA in differing ways. Additional resources have been added in (1) personnel, (2) testing facilities, and (3) recordkeeping/surveillance systems. Other effects have been increased testing, impacts of banned substances and a variety of new or increased internal and external activities.

It is difficult to isolate concise numbers of personnel added since very few are employed full-time for TSCA; they have other duties which flow, directly or indirectly, from related health/environmental/occupational health laws. One area of staff support on which TSCA may have had a sizeable quantitative impact is in toxicologists, who appear to have increased in numbers on corporate payrolls with the implementation of TSCA. Although the volume of testing has increased as a result of TSCA requirements, many of the other related acts require testing. Since much of the testing even prior to TSCA was done outside company testing facilities, the increase in testing volume is to be seen in outside testing facilities rather than in company in-house facilities. New management functions are at least partially due to TSCA. Recordkeeping and surveillance systems development has received a boost from TSCA. A variety of systems ranging from simple, manual to complex, centralized corporate computerized systems exist. Further pressure for such systems may increase as TSCA is implemented.

In general, the larger corporations respond to TSCA demands much as they would to any other demand presented to them. There have been additional resources added; the additional costs will be expressed in the cost of products and services provided by the corporation. With smaller companies, the general result has been to add the TSCA burdens to existing personnel, particularly in technical/research and development functions. The final result in these smaller companies may be less innovation and productivity. In the smallest companies, the burdens have fallen on the few managers whose time in general management functions is reduced. Less productivity may be the result here.

So far as newly created corporate responses, the new management functions (health and environmental Vice Presidents) and additional toxicology expertise are most apparent.

RECEIVED November 22, 1982

Confidentiality of Chemical Identities

JAMES T. O'REILLY

Procter & Gamble Company, Cincinnati, OH 45202

The chemical industry's concern for
confidentiality of innovative chemical products is
a world-wide problem, with a variety of responses
by governments, interest groups and companies.
The EEC system poses the most immediate
confidentiality loss potential. A proposal is
made for a system which excludes certain
identifiable confidential data from the disclosure
provisions of the EINECS inventory.

One day in February 1981, two U.S. government food and drug
inspectors arrived in Shanghai, China, to conduct an inspection
of the Peoples' Fourth Pharmaceutical Works, an enterprise
owned by the Chinese government. It seems that even the
traditionally secretive Chinese were well aware of a distinctly
American phenomenon. When the U.S. FDA inspectors started to
ask questions, the factory manager announced that the Fourth
Pharmaceutical Works was greatly concerned about the
disclosure of their trade secret information to drug makers in
other countries, because of the U.S. Freedom of Information
Act! It's funny how word of this peculiar American pattern of
disclosing private confidential data can create a world full of
skeptics, wondering if the U.S. government can ever keep a
commercial secret . . . but it's frankly <u>not</u> funny that our
government's reputation for leaking innovative technology
information has spread so far and so wide as to reach Shanghai.
This paper offers a perspective on confidential chemical
information, including the identities of existing and new
chemical substances, which are valuable pieces of knowledge
treated by their owners as confidential business information.
The secrets that I will discuss are in three essentially
separable catgories -- formulation secrets, including the
identity of a new special chemical entity like a photographic
film chemical; technology secrets, including the methods and

0097-6156/83/0213-0133$06.00/0

machinery involved in assembling an efficient manufacturing system, and business secrets, like a marketing plan for a new adhesive or the timing of an introduction for a new consumer cleaning prodct. These formulation, technology and business secrets can be treated as typical confidential business information which government agencies obtain, for whatever their purpose, from time to time.

Confidentiality is really not the opposite of public dissemination. Confidential status can have all shades of seriousness; public dissemination has only one. Something can be confidential and vigorously defended, or merely confidential as a matter of common practice. But once the secret is out the bird is on the wing, the arrow is fired, the secret is gone forever. So programs and options for confidential treatment are considered in light of the knowledge that loss of confidential status is permanent. We are "playing for keeps" when we debate whether information should be disclosed, for disclosure alters the existing expectation forever.

That concept of expectation is easy to understand. As a bench chemist working with catalysts, you may have an expectation that you will find the answer to the production dilemma. You also expect to be paid for it in some way. If your grant expires or your company closes before you find the solution, your expectation is defeated. In the same way, our rewards system is designed to allow you to recoup a reasonable license fee from those who want to use your invention, through patent licenses if it is patentable and through trade secret licenses if it is not. You expect to use information to gain academic recognition or peer rewards or financial rewards or some combination of all three. Information puts your work into the form from which its expectation of rewards will flow. You should safeguard the information; you will need it to get the rewards. If you discard the safeguards and that discovery is public, then that novel technology returns no rewards, and the disclosed secret will have filled none of the expectations.

Assume that the bench chemist in the Midwest or Southwest develops a complex new molecule for some use like paint adhesion, plastics color enhancement, or any one of thousands of worthwhile uses. The information can be audited by the government if the program of development is subject to some of the Environmental Protection Agency or Food and Drug Administration regulations governing testing, or if the laboratories where the study is underway have some obligation to permit inspection of their methods as a background for their drug or pesticide studies, for example.

The chemical secret may be in the chemist's changes to the molecule, her process development efforts, or her new and successful application of an old technology. The information which records the success is secret; a patent filing would be possible but may not be economically feasible. At this point,

the expectation is that the secret will remain secret. The
expectation is normally fulfilled; we as consumers benefit from
instant film in brighter colors, or plastics which are less
likely to break. Historically, the innovator has normally
obtained a reward.

But the chemical world faces a kind of cultural
revolution. This is a rapidly-growing political movement to
devalue innovative confidential data. The movement would
overthrow the chemist's expectation of value, in the name of
the social scientist's view of appropriate government
interaction with governmental constituencies. Assume that a
federal agency holds a copy of the confidential data. The
federal government disclosure of that secret would take away
the private person's expectation in one of three ways. First,
the Freedom of Information Act limits the owner's ability to
protect that secret if another person demands disclosure.
Second, the possessing agency has many incentives to disclose
that private business data under the Freedom of Information Act
and no real legal disincentives. And third, the Toxic
Substances Control Act and the Clean Air Act, among others, are
current federal laws which mandate certain disclosures and
force the agencies to make disclosure of what had always been
private information before.

Until chemists become more aware of their information
problem, chemists will be far behind the social science
architects of policies. On openness, protections of technology
which chemists enjoy today were won by legislative battles and
by hard-fought cases under trade secret law which eventually
came to favor the owner. Chemists have to be willing to get
into the political arena and fight for their expectations of
confidentiality, or else be willing to accept the difficulties
of protection which the law now imposes upon innovators in our
society.

I mentioned three factors in federal law which impact on
this disclosure problem. The first is that the Freedom of
Information Act makes it difficult to withhold information. It
does this both by an uncertain standard of confidentiality, and
by the omission of the procedural protection which comes with
all other kinds of adjudicative decisions by federal agencies.
The courts rewrote the Freedom of Information Act's original
intent in the 1974 National Parks v. Morton decision. Since
that time, each submission to an agency has been vulnerable to
disclosure if the owner fails to carry a rather difficult
burden of proof: That disclosure would cause substantial harm
to competitive position at the time the disclosure is made.
Assume that the owner of a secret catalyst had a market share
of 10 in specialty fatty acids for rubber production, and filed
the catalyst information with the EPA on April 1, 1982. When
the request for disclosure comes in November 1983, what will
the firm's market position be then, and how much would this

disclosure hurt that position? It is troubling me to have to
prove that kind of case. Procedurally, the firm is in a quite
uncomfortable position if it does not get notice of the
disclosure--for the law does not require notice, and agencies
vary. While FDA refuses advance notice, EPA often allows
notice to be given. The owner also loses if it does not have a
chance to put that information rapidly into the record for full
consideration by the agency or the courts. The procedures are
rapid; they stress quick decisions. Many chemical firms
concerned with TSCA problems are concerned about the degree to
which courts can become involved in fleshing out the reasons
for the withholding.

The second reality for the chemist who studies federal
disclosure policy is that many incentives favor disclosure and
few favor withholding. A $25,000 a year employee who discloses
chemical process data to a competitor in Ohio can get a year in
jail and a tremendous fine, in addition to damages against the
recipient firm which collaborates with the disloyalty. A
$25,000 a year federal official who discloses the same
information to the same competitor has helped his or her career
in five respects. First, the competitor is pleased and will
remember the cooperation. Second, the public openness
incentives are fostered and the performance ratings will
benefit, for those officials whose bonus can be based on
avoiding conflicts and speeding the process of disclosures.
Third, the employee will face no criminal prosecution since
enforcement of the Trade Secrets Act was effectively suspended
by the Justice Department in 1979 and no prosecutions have
occurred under it. Fourth, the individual employee has no
liability for job-related actions and the agency is exempt from
liability for intentional torts like this disclosure. And
finally, even if the agency cared to rebuke the disclosing
person, discipline is likely to consist of a written reprimand
in the file, as prescribed by such manuals as the EPA TSCA
security manual, and the employee can file grievances or
Privacy Act suits to erase or block even that tiny sanction.
While the former employee chemist is making license plates or
making restitution, the current and future government employee
emerged from the identical set of facts virtually unscathed.
So incentives toward disclosure are more rewarding than any
incentive to withhold.

The final element of this federal impact on confidentiality
is that certain laws require certain things to be made public.
The Toxic Substances Control Act has several useful
confidentiality provisions in §14, but it adversely affects
confidentiality when it prohibits the EPA withholding of health
and safety "studies." That term is not well defined at all.
There are conflicting views about which items are "studies."
From a careful reexamination of the law, I feel that identity
can be confidential and need not be part of a study, if the

chemical name has been excised from study reports. Separately, emission data cannot be withheld under the Clean Air Act; but is "emission data" specific processes within a factory, or complex formulations, or what? No clear definition of what is not emission data can be found.

Beyond these three factors in federal law, the chemical innovator will also face a myriad of state and local rules about disclosure which seem to proliferate in length, in inverse proportion to the   logical basis.

And literally beyond these Washington concerns is the European inventory concern, a matter which many eminent minds in the chemical regulatory field are pondering. A chemical existing in the United States, identified only generically in the U.S. EPA inventory, and if on the EEC market by September 18, 1981, must be reported in terms of its precise identity to the proper authorities in the member state(s) where manufactured or imported for inclusion in the European Inventory of Existing Chemical Substances (EINECS). As the reporting instructions "I. Introduction" Page 7, indicate, "however, no substance identity will be considered confidential in any way." The chemical which has not been manufactured, is not on the U.S. EPA inventory, and which remains a trade secret in the possession of its developer, is placed in jeopardy when commercialized in Europe. Before placing it on the European market, the U.S. manufacturer representative in one or more of the member states of the European Economic Community must submit a pre-marketing notification to the authorities in the member state(s) of choice. No more than three years of protection is available, and that three year limit is contingent upon the material's being classified as not dangerous.

A conflict clearly exists between permanent confidentiality, available under the system of U.S. laws, and the eventual disclosure of identities of specialized chemical substances which had heretofore been undisclosed, but which are now affected by EINECS or by EEC's premarketing notification system. The rules are different; the assumptions regarding disclosure are different. Perhaps the best solution a lawyer could offer is that member states should be willing to adjudicate individual cases of specific confidentiality needs. Inventories of existing substances are rules, adopted prospectively to announce to the world both the existence of a material and its regulatory status. Those rules can operate to accommodate both public and private needs.

We have a U.S. sytem which could satisfactorily resolve the conflict. I propose that we take the approach of permitting confidentiality claims for the identity of an existing substance, if they can be justified to the national authority, such as a ministry of the environment, which will consider whether confidentiality reasons are sufficient to merit excepting that item from inventory requirements.

The national authority which is processing the information
for EINECS could refuse the claim, in which case the firm must
either allow for disclosure or use available remedies to appeal
-- or not use the material within the EEC in order to preserve
confidential status elsewhere. Or the national authority could
accept confidentiality, and could insist that it be told
immediately upon public disclosure (or patent issuance
disclosure) of the material's existence.

The system is modeled on the U.S. Food and Drug
Administration cosmetic trade secrecy system. That system
predates the enactment of TSCA, but it offers a relevant
benchmark for protection of confidentiality. The EPA and other
agencies can lawfully permit just and equitable exceptions and
exemptions from its rules on a basis of specific facts. Under
FDA's system, all cosmetics must disclose ingredients on their
labels, except those which have used the FDA procedure to win a
letter acknowledging the specific ingredient's specific
confidential status. A roughly analogous situation to the
inventory occurred in 1958 when Congress adopted the Food
Additive Amendments and did not wish to have all existing
substances reexamined; Congress allowed for prior specific
sanction letters by FDA to excuse the maker of the additive
from submission of a food additive petition. It would be best
for the EEC to offer the same form of special letter
determinations to submitters of confidential data. Then the
submitter who wanted to comply, but would lose its U.S. and
other non-EEC confidential status through disclosure, would be
covered. Those who felt there was a legitimate reason for
nondisclosure could ask the agency for official approbation for
them to do so.

The flaws in the FDA's systems have been twofold. First,
the cosmetic secrets system has not given adequate
consideration to private statements showing confidential
status. Second, the central agency staff must know how many
exceptions have been allowed, to whom and for what. If the
procedure is unfair or if exceptions cannot be catalogued, then
the agency cannot manage its workload and the submitter cannot
be adequately listened to. But these problems can be worked
out within the EEC structure, and chemists familiar with the
EEC inventory process should aid in the letter-exceptions
process.

The flaws in the U.S. system, particularly in the Freedom
of Information Act, are not easy to fix. Legislation now
pending in Congress will resolve the procedural flaws. It
will, for the first time in eight years, protect academic
research, which has not been protected by the law since
government has not considered it "commercial." Changes to
specific laws like the Pesticides Act, FIFRA, have been so hard
to effectuate that we see that TSCA and the other statutes will

be quite difficult to reshape into any more appropriate
confidential format. Perhaps administrative changes, or the
new Office of Management & Budget authorities under the
Paperwork Reduction Act, will provide the basis for legal
change.

These debates about product identity are separate from the
debates about technology secrets and business secrets. It is
often quite difficult to convince a federal agency that a piece
of business information will do harm if disclosed. The EPA has
been one of the most enterprising and accomodating when the
details of specific disclosure consequences have been
explained. But where a prodisclosure policy intervenes rather
than the consideration of a specific set of facts, that policy
is often flawed by a factual or legal problem. The court
decisions in several recent cases indicate that a well-grounded
confidentiality protection program can succeed. Technology can
be protected, but more work needs to be done. Business secrets
need to be connected to competitive marketplace disadvantages,
as the current standard requires, and that can be quite complex
as a matter of proof.

The future problems with confidentiality of chemical
industry innovation will be significantly different from
problems experienced in the past. At one time, the laws of
another country dealing with trade secrets were only of
interest to the special breed of corporate lawyers who
prosecuted multi-national conspiracies to steal secrets through
industrial espionage. Today, by constrast, the planners of
multi-national enterprises must pay attention to the
requirements of other countries for disclosure. Where a
Pennsylvania company may wish to protect its identifiable new
chemical from disclosure in Belgian markets, the Belgian firm
making the current chemical for a new use may be concerned
about sending it into Pennsylvania because of the uncertainties
about commercial data protection of that new application or new
use, as a matter of U.S. governmental disclosure policies.

My predictions would be for legislated changes, though
legislation takes more effort, more money and more coordination
than most of the alternative but weaker options. The
legislative changes will define the interests to be protected
and the ones that will not be protected. It will require some
information to be disclosed where there is a real public
interest, one which overrides the private interests at stake.
On the European front, there will be some accommodation of the
conflicts, perhaps through an adjudicative mechanism for
specific products that does not depend on the central listing
device of the Sixth Amendment inventory.

The missing variable in these predictions is time. Until
there is a change in the European inventory provisions there
will be an exposure of the confidential identities immediately
upon publication (expected in 1984 or 1985) for existing

chemicals. For non-dangerous new chemicals this disclosure
will occur three years after Notification to the Member State
Competent Authorities. This change in the Inventory provisions
needs to be worked on promptly so that the U.S. and European
Confidentiality Provisions can be compatible. Freedom of
Information Act reforms are being worked on in the Congress,
but reform is slow and always uncertain. The EPA has operated
under the same set of confidential information rules for six
years now, and the time for reexamination of its regulatory
requirements for information submitters is coming due. The
increase in EPA Premanufacture Filings, year by year, increases
the amount of confidential business information that chemical
firms are sharing with the Government and about which
protective provisions will become increasingly important.

Meanwhile, chemical innovation goes on. Expectations are
set, confidences shared, submissions to Government made, and
work proceeds normally with research and development. A new
social policy favoring disclosure may be in the works, while
somewhere in China a factory manager is wondering what
mysterious measures the Western Countries will come up with
next. This worldwide concern will be with us for years to come.

RECEIVED August 18, 1982

Impact on the Reactive Polymer Industry

STANLEY C. OSLOSKY and LAWRENCE W. KELLER

PPG Industries, Inc., Springdale, PA 15144

This paper discusses the impacts that the Toxic Substances Control Act (TSCA) has had on the reactive polymer industry. Characterized by the need for timely introduction of innovative products to satisfy a constantly changing market, this segment of the chemical industry has been affected more than any other. Five years of experience with TSCA in general and nearly two and one half years with Premanufacture Notification requirements have produced some positive and some negative impacts, particularly at the research and development level. On the positive side, the Act has forced criteria for improved planning, opened some communication lines and strengthened others, created data bases that can aid in research and development, and generally encouraged a broadened staff perspective from the economic and technical toward a total product awareness. On the negative, certain aspects of the implementation of the Act have inhibited innovation, caused necessary duplication of information submissions, and have created general uncertainties by inconsistencies in data treatment and the lack of a finalized policy.

The recent economic legitimacy which is given industrial concerns over the impacts of government regulations brings sharp focus to the impacts that the Toxic Substances Control Act (TSCA) has had on innovation. Reactive polymers are a major component in many products in segments of industry that can be characterized by the need for innovative products to meet constantly changing market needs. Such products include

0097-6156/83/0213-0141$06.00/0

coatings, plastics, inks and adhesives. The essential role of these substances in the economy is of little question. They, in some way touch nearly every part of the Gross National Product.(1) Their ability to be engineered at the molecular level gives them the ability to provide enhancement, protection and low cost, lightweight alternatives to many time-honored materials.(2) It is this versatility that makes them particularly subject to regulations such as TSCA.

The Center for Policy Alternatives at the Massachusetts Institute of Technology, under contract to the EPA, observed that regulation may affect industry in areas such as profitability, growth, imports, exports, employment and technological innovation.(3) From the Coatings and Resins perspective, direct impacts on innovation have had a cascading effect on the other areas with the exception of direct employment costs associated with compliance.

Congress recognized the potentially detrimental effects that TSCA could have on innovation while recognizing the need to assess each new substance entering the market for potentially adverse effects. This is evidenced by the often quoted Section 2(B)(3) that says the agency's authority: "... should be exercised in such a manner as not to impede unduly or create unnecessary economic barriers to technology and innovation while fulfilling the primary purpose of this Act to assure that such innovation and commerce in such chemical substances and mixtures do not present an unreasonable risk of injury to health or the environment."(4)

The regulations that have been troublesome to innovative industries have been promulgated under the authority of Sections 5 and 8 of the Act. Initially, industries met promulgations under subsection 8(b) which deals with reporting and retention of information. This required submissions geared toward compilation of an inventory of chemical substances commercial in the United States. Also, since TSCA's enactment, subsection 8(e) requirements have been in effect and require reports to the Administrator of all "substantial risks", defined as adverse effects on environment and health. To comply with 8(e) industry has to provide avenues of employee input into the compliance process. Regulations have been in proposed form under subsections 8(a) and 8(d). If finalized, regulations promulgated in these areas would impose further informational submission requirements on certain inventoried substances.

While Section 8 requirements usually created some significant temporary impacts, Premanufacture Notificaton has had effects that require industries whose profitability is based on innovation to make some changes in their business operations, particularly at the research and development level.

Aside from the Premanufacture Notification (PMN) requirements, Section 5 rules have not been used to a great extent.   Nearly all of the compliance activity since July of 1979 has been geared toward meeting PMN requirements.   Whether the agency's failure to use the subsections of 5 that are designed to restrict potentially harmful substances from reaching the marketplace has resulted from lack of real necessity, a failure in the Agency's risk assessment process, or an inherent non-workability in the Law itself, remains to be seen.

If industrial activity or the amount of effort and resources directed toward compliance could be measured and plotted against time, it would look something like Figure 1. From TSCA's enactment in 1976, there was a slow, gradual rise in activity, which reached a peak in the first quarter of 1978, with the first 8(b) reporting deadline.   A relaxation was then experienced, a gradual rise and a second peak corresponding with the official publication of 8(b) inventory in 1979.   A decline in TSCA related activity was then seen during a period when only a very onerous proposed PMN regulation guided those wishing to enter a new substance into commerce.   PMN activity slowly picked up pace as a more reasonable proposal was won and the process became learned.   Activity, geared toward compliance, is now most likely in the process of declining and leveling above some pre-TSCA baseline.   Definitions of this baseline and quantification of the impacts have been the subject of several extensive studies(5, 6, 7) and have proven to be controversial.   With compliance programs firmly established, and given the status quo of regulatory initiatives (e.g. toward requirements relaxation), a definition of some of the positive and negative components of the impacts that have caused deviations from this elusive baseline can be constructed.

Negative Components

Negative components of the impacts of TSCA can be divided into three general areas, resource diversion, testing costs and uncertainties.   Positive impacts offsetting the negative can be divided into four components;   resource diversification, strenthening of communication lines, improved planning and a general redirection of technology toward safer and more healthful products.[3]

Resource   Diversion.   First in the negative area is resource diversion or the diverting of facilities, costs and personnel from their characteristic functions toward some compliance effort.   One of the reasons that quantification of impacts associated with TSCA is difficult, is that the

compliance accountability has become dispersed throughout the
research and development process.   Initially,   diversion
occurred in the form of upper and middle management
informational and planning sessions geared toward designing
"required" compliance measures. These eventually evolved into
a product information flow in which very little product
information remained free from input.   Effective ways of
assessing   production   histories,   grouping   compositional
duplicates, mapping production distributions, substantiating
commerciality and filing site-specific 8(b) reports for
compositions had to be devised.   A retrospective information
flow was established to a coordination center, usually becoming
a function of newly emerging Health and Environmental Affairs
Departments, and the results were a mass of 8(b) reports filed
for every recently commercially valuable chemical entity, from
every site of manufacture. The magnitude of the task in the
polymer industry required a significant diversion of personnel
and other resources through 1977 to May, 1978.

     After the initial inventory reporting cutoff in May of
1978, this retrospective approach was replaced with the flow
of current information.   New substances, as they became
commercial, had to be identified and put through a decision
process as shown in Figure 2.   New reports were again
necessitated by the commercialization of a new substance.
Since July of 1979, Section 5 Compliance has been the source of
nearly all compliance activity.   Input for Premanufacture
Notification purposes now takes a shape similar to that
previously indicated, (see Figure 2) but now incoming new
product material must reach the coordinator earlier in the life
of the product than at any other time in the compliance effort.
This introduced a potential time delay in the commercialization
process.

     Where in a product's life a regulation such as TSCA is
imposed becomes important in terms of resource diversion.
Ideally, imposition of Premanufacture Notification after
scale-up and successful line trials would necessitate filing
notifications only for those products that actually become
commercial. In the usually compressed Coatings and Resins
lifeline, this is not always, and in fact very seldom,
possible.

     Because of strict definitions of commerciality,   PMN
initiation must occur so early to allow sufficient time for
internal formalization and ninety days of EPA review that many
of those going through the review process will be altered due
to continuous innovation efforts to improve performance and
cost. These alternatives often result in nearly structurally
identical substances but ones that TSCA will define as "new".
Even when these products do reach commerciality, the same
research effort often offers them a very limited life span.

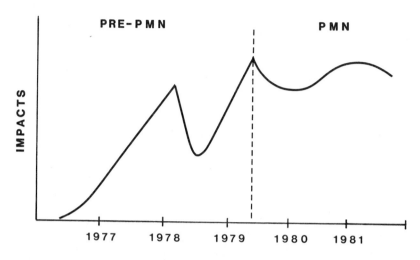

Figure 1.   Perceived TSCA impacts.

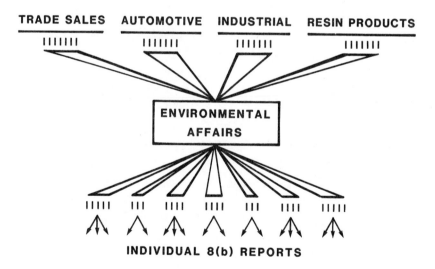

Figure 2.   Original compliance information flow.

When the amount of effort associated with each PMN and the amount of PMN's filed are considered, it can be seen that total diversion can become significant. A typical Coatings manufacturer's experience during two full years of PMN compliance has shown that a synthetic chemist in the Coatings and Resins area brings an average of three substances per year into the notification process. This is reduced to approximately two when considering just substances brought to the point of actual submission. These are documentable associations, however, much of an innovative chemist's concern occurs in activity difficult to document, such as maintaining an awareness of the inventory status of all substances synthesized in a particular area, so that effective initiation decisions can be made when feedback from product development areas dictate. Documentable resource diversion, at the research level, consists of gathering and formalizing necessary data and participation in review meetings. Other levels of input, development, processing or scale up and Environmental Affairs, have similar input requirements and can add to make the diversion of personnel per Premanufacture Notification quite significant. This is before many other factors which add to this are even considered. Other considerations are time spent in activities such as: coordination and requirements consultation; formalization and record maintenance; data base maintenance and other computer related activities such as inventory and literature searches; and finally, upper management reviews of notifications before submission. In the short term, TSCA requirements divert resources at the research and development level toward compliance efforts and away from innovation.

Chemical manufacturers submitted 1,031 Premanufacture Notifications in 1980 and 1981. In the same time period, they submitted 290 notifications of commencement of commercial manufacture. In other words, only 28% of the substances for which Premanufacturing notices were filed in the past two years of compliance have become commercial. Specifically, for the reactive polymer segment, about 29% of the reported substances have become commercial. These percentages indicate that much of what has constituted the impacts of TSCA has been "protective" filing of notifications. The length and complexity of the process mandated by TSCA has led to unnecessary resource diversion. A more liberal definition of what constitutes a commercial event could have significantly increased the percentages and reduced this type of impact, at least since July, 1979.

The possibility exists that the ratio of Commencement Notifications to PMNs submitted will continue to increase. If

the trends are considered and the number of PMNs submitted per
month for the years 1980 and 1981 are plotted using a three
month moving average, a steady upward trend can be seen (see
Figure 3) with slight drops around the end of 1980 and the
summer of 1981.   The lower area shows the coooresponding
averaged Commencement Notification submissions.   Commencement
Notification trends as a percent of total PMN submission can be
plotted (see Figure 4) and appear to be leveling around 40%.
If this trend continues, over 60% of the PMNs submitted will be
for substances that never reach commerciality.   The ideal
situation would be Commencement Notifications overtaking PMNs
but at least an equalization would be welcomed in terms of
diverted resources.

Testing Costs. The second negative area is the cost of
testing that is done to assess the toxicity of new substances
entering commerce. This would seem to be a more quantifiable
negative impact and in industries where a time lapse between
line trial and scale-up exists, this is likely the case.
Toxicological testing would be imposed somewhere in that period
and any additional testing done for compliance purposes is
easily separated from what would have been normal testing.   As
discussed, this becomes less possible as competitiveness of the
market areas increases. Manufacturers in fiercely competitive
industries, where the commercialization process is compressed,
are forced even further back along the product
commercialization lifeline to allow development of supporting
data, sufficiently anticipating PMN submissions. Forcing early
risk considerations and decisions is not an entirely negative
effect, but if a closer look is taken at the way business is
done, it can be seen that the potential for much unnecessary
testing exists even in commercially successful product lines.
In polymer industries increasing occupational exposures
to a decreasing number of products tend generally to occur as
commercialization is approached. Through basic resin research,
development activities, scale up and line trials, an increasing
amount of exposure is typically experienced. Commercialization
can also bring exposures that are quite significant depending
upon the particular end use. It can also be said that the
ratio of individuals exposed, who are not "technically
qualified"(8) to recognize potentially adverse effects,
increases similarly. The nature of the product throughout this
period also changes. There are seldom exposures to other than
a mixture containing the new substance and this mixture becomes
increasingly complicated. As a potential product is brought
toward commercialization, more employees, primarily
non-technical, will be exposed and the substance to which they
are exposed will look substantially less like the "new

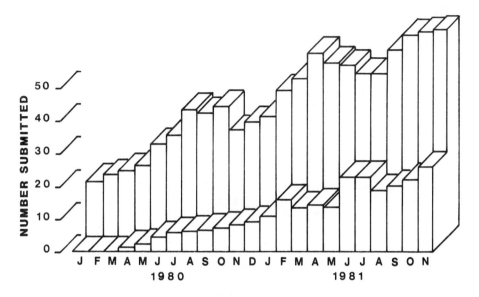

Figure 3.   Three-month-averaged PMNs (upper area) and commencement notifications (lower area).

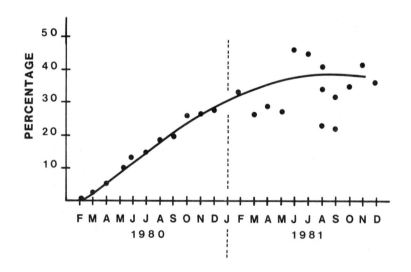

Figure 4.   Commencement notification as percent of PMN.

substance" identified in the Premanufacture Notification form. In this setting, any action which forces testing decisions earlier in the research and development life of the product will:

1. increase the incidence of testing products which ultimately fail;
2. develop information on pure substances with relatively limited exposure;
3. direct valuable resources from tests on final formulations which have much higher exposures.

PMN requirements have had these effects on the coatings polymer industry. There are some long-term positive aspects which will be discussed later.

In order to assess the amount of testing that was done for reactive polymer type materials an extensive Federal Register submission summary review was conducted. Selection criteria were: those submissions attributed directly to a coating or resin end use; those submissions for materials directly related to a coating and resin end use, such as paint pigments and polymerization catalysts; and those submissions, representing 16% of the total submissions in both years, which were so confidential that some subjectivity was necessary in assigning them by generic name to a particular group. This review is summarized in Table I. During this review, all submissions were also assessed in terms of toxicological data content. It can then be seen in Table II that the number of PMNs submitted as coatings and related materials and containing at least one substance-specific toxicological test, rose from 16% in 1980, to 30% in 1981. Similarly derived unrelated submission percentages dropped from 22% to 15% over the same period. Remembering from Table I, that for both years substances in coatings and related categories represented slightly more than half of the total submissions, it can be seen that testing initiatives for products in this area doubled in the second full year of compliance.

Uncertainty. The final component of negative impact is uncertainty. To say that there has been some industrial uncertainty during the implementation of TSCA is certainly an understatement. Indeed there are still pending; finalized Premanufacture regulations; Section 8(a) and 8(d) proposed regulations that have been in a constant state of uncertainty; Significant New Use Rules (SNURS); a constantly growing list of Interagency Testing Committee recommendations, and the individual new product uncertainty that each manufacturer faces every time the ninety day review period must be faced.

Not knowing how a potentially promising product will fare in the regulatory arena often prompts premature action.

Table I.   Percent of PMN Submissions For Coatings, Resins, and
Other Related Substances

	1980	1981
C&R PMN'S	37%	33%
C&R RELATED PMN'S	4%	5%
POSSIBLE C&R (FROM CONFIDENTIAL)	16%	16%
TOTAL	57%	54%

Table II.   Percent of Total PMN Submissions with Tests

	1980	1981
TOTAL PMN WITH TOXICOLOGY	38%	45%
NON C&R TESTS	22%	15%
C&R TESTS	16%	30%

Looking at a typical compliance experience illustrates this. In Figure 5, the number of submissions of one manufacturer are graphed for each compliance year, as a percent of their total overall submissions for all four years. A very low 6% in 1979, and 3% during January and February of 1982 can be seen. If projections are made for 1982, based on an average of the percent of the total that were submitted during the first two months of 1980 and 1981, submissions this year should only reach 12% of their total. PMN regulations were in effect for the last half of 1979 and the low figure shown is likely to be the result of six months of resources diverted in an attempt to accelerate commercialization to avoid the uncertainties of the upcoming regulations. At that time, only the original January 10, 1979, version of the proposed regulations(9) had been published and these are several orders of magnitude more cumbersome than anyone had anticipated. Accelerated commercialization efforts allowed an industrial regrouping until a more reasonable proposal was won.(10) Submissions then began to grow rapidly. Throughout the PMN compliance period, and particularly in 1981, it became evident that the notification process was not as onerous as most first feared. Current developments portend even further relief for polymers and as this information filters out, a decline in the number of PMNs filed in 1982 will no doubt be seen. Manufacturers will be more confident about the ability of their substances to withstand the process and will hold off filing until commerciality is more certain. Although a decline is being seen in a typical compliance example, it is not being seen as yet nationally. Figure 6 shows comparative January and February receipts as surrogates for annual compliance data for '80 and '81; a surprisingly consistent rate of growth can be seen.

Uncertainty is a factor in all of the impacts associated with TSCA. It was seen that it is a source of unnecessary resource diversion and toxicological testing. How much of a factor it becomes is a matter for debate. It can be said that for an industry such as coatings polymers, it is related to the company's commitment to safe and healthful products in the absence of TSCA. The more of a safety and health commitment that existed pre-TSCA, in that innovation was strongly linked to such concerns, the less uncertainty is going to be a significant impact.

Positive Components

Discussion of the negative impacts experienced with TSCA made several allusions to the long-term positive aspects of these effects. Where difficulty was experienced in quantification of negative impacts, even more difficulty is experienced with the

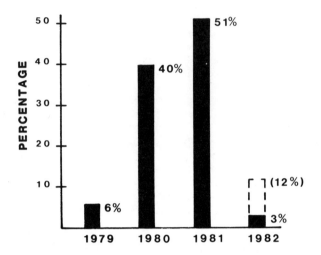

Figure 5.   Typical compliance experience as percent of total PMNs.

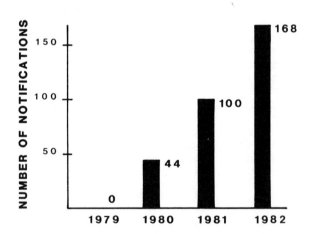

Figure 6.   PMNs received in Jan. and Feb.

positive. Three general effects will be seen; resource
diversification, or the increase in the ability of existing
personnel and products to minimize compliance impacts; the
establishment of new and improvement of old communication
lines; and finally, improved planning.

Diversification. First diversification occurs because
resources are diverted toward less characteristic functions.
Two specific areas where TSCA has had impacts are personnel and
products. Research, development, manufacture, process and
Industrial Hygiene personnel must meet early in the potential
product's life to make predictions concerning areas specific to
each. This tends to give anyone present a somewhat broader
view of the products with which they are involved.
Product diversification, defined as the ability of
existing products to meet new needs, also becomes a positive
component. Since TSCA makes it advantageous to find existing
products to meet new market needs and avoid the notification
process, resources diverted toward compiling data bases which
had previously grouped chemicals by structure can now be used
to that more suitable, positive end use. The potential exists
in programs such as these for a great deal of sophistication
and value in formulation. Access can be through variables such
as property or toxicological ranges; trends and usages can be
tracked; and an easily accessible, centralized source of
information on produced substances is almost compulsory in
terms of emergency response and product liability. So if the
generation of less products to meet the same needs is
advantageous, then some of the early diversion of technical
personnel toward compliance activities could be eventially
construed in a more positive light.

Communication and Planning. Next, positive effects can be
seen when TSCA is imposed on the commercialization process in
terms of communication. New lines are opened in terms of
communications between research and development personnel and
those associated with health and environment. As discussed
before, with actions forced early in product initiatives,
expedient contact between groups whose contacts were previously
limited to post-commercialization are necessary.
Pre-commercialization communication in many cases, took place
only if some property or exposure situation precluded
commercialization without toxicological considerations.
Earlier communication also has the positive effect of orienting
decisions concerning such preclusions toward appropriate
expertise. A strengthening of existing communication lines
within research and development can also be seen. Expedient
feedback as the product approaches marketing is made necessary

on a formalized basis to ensure that the most appropriate compliance decisions are made.

Not much can be said of impacts on planning except that improved planning is not a bad idea in any endeavor and that TSCA compliance urges improvement at all levels.

In summary, the Toxic Substances Control Act has, as expected, impacted significantly on the innovative process in the Coatings and Resins Industry. If viewed purely as another burdensome regulation, the negative impacts can far outweigh any benefits. Nearly every negative impact discussed has some positive aspect that the manufacturer committed to health and safety can maximize to reduce the total adverse effects. With new EPA priorities, designed to lessen the impacts on innovative industries, and an increased commitment in the polymer industry toward testing, a final net positive redirection of technology toward safer and more healthful products is not an impossibility.

Addendum

Though the rate of submission of Premanufacture Notifications is expected to decline, the number of risk assessments that must be conducted every month is unreasonable for any central body. It was seen previously that, in general, polymer manufacturers are becoming more committed to assessment of the toxicological properties of their materials and this is probably true of product evaluation in general. With the current reductions of agency resources, some of the duplicated effort could be eliminated by allowing, through the exemption pathway, industrial risk assessments to be conducted. In effect, some of the proposed exemptions that are currently being considered cursorily accomplish this. These could be expanded and the risk assessments themselves could be approved by the agency with submission contingent on the substances' failure of this internal system. Companies now use this type of system to make submission decisions, and the low percentage of materials that have failed the EPA assessment process is perhaps an indication that such a system could work.

Literature Cited

1. Dean, John C. Chemical Week, October 21, 1981.
2. Shreve, R. Norris; Brink, Joseph A. Jr. "Chemical Process Industries"; McGraw-Hill Inc.: New York 1977, p 571.
3. Anon. "Supporting Innovation: A Policy Study", Center for Policy Alternatives, Massachusetts Institute of Technology (September, 1980).
4. "The Toxic Substances Control Act", Public Law 94-469, October 11, 1976).

5. Little, Arthur D., Inc., "Impact of TSCA Proposed Premanufacturing Notification Requirements", Office of Planning and Evaluation, U.S. Environmental Protection Agency (December, 1978).
6. ICF, Incorporated, " Economic Impact Analysis of Proposed Section 5 Notice Requirements", Economics and Technology Division, Environmental Protection Agency (September, 1980).
7. Heiden, Edward J.; Pittaway, Alan R. "A Critique of the EPA, Economic Impact of Analysis of Proposed Section 5 Notice Requirements", Chemical Manufacturers Association (March 16, 1981).
8. Anon. Toxic Substances Control: Inventory Reporting Requirements FR. 1977, 42, 46576.
9. Anon. Toxic Substances Control: Premanufacture Notification Requirement and Review Procedures FR. 1979, 44, 2241-2348.
10. Anon. Reproposal of Toxic Substances Control Act (TSCA) Premanufacture Notice (PMN) Forms and Previsions of Rules FR. 1979, 44, 59763-882.

RECEIVED September 29, 1982

Effects of TSCA on the Metalworking Fluids Industry: Increased Awareness of Nitrosamine Contamination

HOWARD M. FRIBUSH

U.S. Environmental Protection Agency, Office of Toxic Substances, Washington, DC 20460

The various effects of the Toxic Substances Control Act (TSCA) on the metalworking fluids industry is presented, with emphasis placed on nitrosamine contamination of the fluids. A review of the literature on the effects of various metalworking fluid additives on nitrosamine formation is also presented to aid the industry in dealing with the nuisance of nitrosamine contamination. It is concluded that with increased awareness of nitrosamine contamination as a result of the implementation of TSCA and careful consideration of the factors described in this paper, it may be possible to design and control a nitrosamine-free metalworking fluid.

The Toxic Substances Control Act (TSCA) was passed by Congress in 1976. Congress enacted TSCA because it felt that the health and environmental risks presented by the production, distribution, use, and disposal of certain chemicals may be unacceptably high, or "unreasonable".

TSCA provides the EPA with a number of specific authorities to gather information and, where necessary, to control unreasonable risks. Section 4 gives EPA the authority to require manufacturers or processors of chemicals to test certain substances for toxic effects. To identify areas of greatest concern, Congress created under this section an interagency testing committee (ITC) composed of eight members drawn from various federal organizations. This committee recommends chemicals to EPA for priority consideration for test rules, and EPA must decide within one year what action, if any, to take on these recommendations.

Although Congress recognized that not all chemicals presented such risks, it had no mechanism for determining which chemicals to target for control prior to enactment. Congress also recognized that a balance would have to be struck between the need for mitigation of serious risks on one hand, and the benefits provided by chemicals and the costs of controlling risks on the other. Section 5(a) requires manufacturers who wish to introduce into commerce a chemical not on the TSCA Inventory to notify EPA at least 90 days before beginning manufacture. Congress thus recognized that the most desirable time to determine the health and environmental effects of a substance occurs before commercial production begins.

Section 8(e) of TSCA requires any person involved in the manufacture, processing, or distribution in commerce of chemicals to immediately notify the Agency of any information on a chemical which supports the conclusion that the chemical presents a substantial risk to health or the environment. Under Section 6, EPA may find that a chemical will present a risk to health or the environment and thus, must be controlled as a hazardous substance. EPA may apply one or more of the following controls to protect against the risk, using the least burdensome requirement and balancing social and economic factors, including benefits and availability of substitutes. EPA may (1) prohibit or limit production, distribution, or the amount handled; (2) prohibit or limit the amount used; (3) require labeling or instructions for handling; (4) require recordkeeping of production processes, monitoring, or testing for compliance; and (5) control disposal.

Metalworking fluids are essential for metalworking as they cool and lubricate cutting machines during the metalworking operation. With machine improvement and increased restrictions on the disposal of oil and grease, these fluids have been modified to minimize disposal problems and to accommodate higher speeds while increasing tool life. This technological achievement has resulted in a shift from oil-based to the rapidly growing water-based fluids, which contain various lubricant additives and preservatives. Nitrosamine formation in water-based metalworking fluids is a problem of which the industry is acutely aware, and has responded to by developing substitutes for the potent nitrosating agent nitrite. Nitrosamine formation in metalworking fluids appears to be affected by several factors, including (1) catalysis by metals, certain biocides, and nitrogen oxides; (2) surfactants; (3) pH; (4) unsuitable primary amines; (5) heat; and (6) storage of concentrates.

It must be noted that the metalworking fluids industry does not represent a unique situation and thus is not being isolated from the various sectors of the chemical industry. Rather, the nuisance of nitrosamine contamination in metalworking fluids has been recognized both by the industry and government for at least six years, and has provided the industry with an opportunity to study nitrosamines and voluntarily reduce or eliminate the problem. In fact, the metalworking industry deserves special recognition for the progress it has made in discovering replacements for nitrite, thus reducing dramatically the levels of nitrosamines in metalworking fluids. This paper attempts to demonstrate how TSCA can be used to aid a particular chemical industry, and to show how government and industry can work together to solve a particular problem—the inadvertant contamination of metalworking fluids with small amounts of nitrosamines. In particular, effects of relevant sections of TSCA on the metalworking fluids industry will be discussed followed by a technical analysis of the effects of the fluids on nitrosamine formation.

Major Trends in the Metalworking Fluids Industry

Section 8 of TSCA has made EPA and the industry aware of nitrosamine contamination in metalworking fluids. In one particular potentially significant notice of substantial risk (8E-1077-0 012), skin painting studies showed an increase above the expected normal incidence of tumors in the livers and lungs of mice with no unusual incidence of skin tumors. Due to the similarity between the observed effects and the mechanism of action of nitrosamines (i.e., the apparent systemic effect of the substance and organ specificity of the tumors), the company attributed the response to a nitrosamine contaminant in the fluid.

With respect to the Agency, Section 5 of TSCA has made EPA more familiar with trends in the metalworking fluids industry, the chemical components of the fluids, and the interactions between the various components. Major trends in the industry are (1) a shift from the traditional oil-based to the rapidly growing water-based fluids; (2) a shift from the use of the nitrosating agent nitrite as a rust inhibitor; (3) the use of multifunctional additives; and (4) the careful monitoring of various factors and additives associated with these fluids.

The shift from straight oils to synthetics has resulted mainly from machine improvement and increased restrictions on the disposal of oil and grease. Thus, metalworking fluids have been modified with water to minimize disposal problems and with chemicals to preserve the fluids and to accommodate higher speeds while increasing tool life.

Typical cutting and grinding fluid concentrates, for example, may contain from 10-20% emulsifier (1), 25-50% lubricating agent, and 0-1% antimicrobial.   They may also contain 1-10% corrosion inhibitor (2), unless the emulsifying and corrosion inhibiting properties are provided by the same additive (3). These formulated concentrates are later diluted with water 10-100 fold for the actual metalworking operation. (4, 5, 6). Although preferences among straight oils, semi-synthetics, and synthetics vary, in general the most recently developed synthetic fluids appear to provide superior performance to the more traditional fluids. This has resulted in an increased demand for synthetics, and a rather large number of chemicals are now available for use in these fluids, with selection of a particular fluid based on a state-of-the-art knowledge.

Unfortunately, not all combinations of chemical additives in water-based fluids are completely compatible, and side reactions leading to various byproducts have been noted. The best known of these side reactions is the reaction between the corrosion inhibitor nitrite and the emulsifiers di- and triethanolamine (7) to form N-nitrosodiethanolamine (NDElA), a nitrosamine reported to have carcinogenic activity (8, 9, 10). In fact, most nitrosamines are carcinogenic, and no animal species which has been tested is resistant to nitrosamine-induced cancer. Although there is no direct evidence that firmly links human cancer to nitrosamines, it is unlikely that humans should be uniquely resistant.

Since nitrosamines have been detected in "certain localities" it is imperative that populations exposed to nitrosamines be identified. (11)

Nitrosamine contamination in metalworking fluids is a problem of which the industry is and has been acutely aware, and has responded to by voluntarily developing substitutes for the potent nitrosating agent nitrite. Receipts by EPA of premanufacture notices (PMN) under Section 5 of TSCA also reflect a high level of research activity by chemical manufacturers to develop corrosion inhibitors for metalworking fluids which will not lead to the formation of nitrosamines during the normal use life of the fluids. For example, in more than one instance a presumably nitrosamine-free nitrite replacement was shown to be contaminated by or potentially form nitrosamines. From our knowledge of this area, we surmised that the nitrosating agent was a byproduct from manufacture of the feedstock used to produce the PMN substance. This assumption was further supported by the technical data sheet describing the feedstock, which indicated that it contained as an impurity roughly 10 % of a nitrosating agent. Subsequent analysis by the manufacturer confirmed the source of the contamination. The problem was solved by substituting a purer form of the feedstock, at a minimal cost increase, which resulted in a considerable drop in the nitrosamine level. In another PMN case the chemical manufacturer produced a fluid that was incompatible with the presence of nitrite, breaking down when nitrite was added. Thus, with respect to the industry the TSCA program may have (1) stimulated a greater awareness of the presence of nitrosamines in metalworking fluids; (2) stimulated a greater awareness of the potential of the various raw materials to give rise to nitrosamines; (3) emphasized the need for nitrite substitutes; and (4) stimulated thought and study to determine the various effects of metalworking fluids on nitrosamine formation.

Nitrite substitutes can be divided into seven chemical categories: (1) amine benzoates; (2) fatty acid amines; (3) phosphate or carbonate silicates; (4) organophosphates; (5) amine borates; (6) alkanolamines; and (7) quaternary ammonium compounds ("quats"). Thus, the technology already exists for replacing nitrite with no loss in rust protection. However, most replacements for nitrite are more expensive, less effective, less likely to be compatible with other additives, and work by a different mechanism (12). It is therefore not surprising that fluids containing nitrite are still relatively common.

Because water-based fluids do not last as long as the more conventional oil-based fluids, careful monitoring of fluids is required. In addition to the standard analyses for pH, dirt and metal fines, dissolved iron, and tramp oil, the introduction of various chemical additives has required the additional monitoring of organic amines, ammonia, rust inhibitors, water hardness, and even nitrosamines in some cases.

The following discussion is intended to show how metalworking fluids can affect nitrosamine formation, even when their formation is not apparent. It is also intended to provide the industry with information that may enable it to cope with unforeseen nitrosamine contamination in metalworking fluids.

History of Nitrosamines in Metalworking Fluids

The issue of nitrosamine contamination in metalworking fluids was a relatively quiet topic until 1976 when four reports appeared in technical journals, magazines, and government bulletins. Zingmark and Rappe (13) showed that NDE1A could be synthesized under simulated gastric conditions from a grinding fluid containing triethanolamine and nitrite. From unpublished, preliminary studies by Fine, the National Institute of Occupational Safety and Health published a Current Intelligence Bulletin which presented industrial hygiene practices that could help reduce dermal and respiratory exposures to metalworking fluids (14). This information subsequently appeared as bulletins in Chemical Week (15) and Chemical and Engineering News (16).

Fine's preliminary monitoring study in the U.S. showed that 100 0 ppm NDE1A was present in a diluted fluid prior to use, and that 40 0 ppm remained after use. Furthermore, in a UK machine shop, 60 0 ppm NDE1A was found in the fluid. Spurred by the results of these preliminary findings, Fine, in a random selection of eight commercially available fluid concentrations in the Boston area, found NDE1A levels ranging from 0.02 to 3% (7). These results strongly indicated that secondary and tertiary amines are precursors to nitrosamines in nitrite-containing metalworking fluids.

NDE1A is not the only nitrosamine reported to have been identified in metalworking fluids. A fluid in the Netherlands was found to be contaminated with 5-methyl-N-nitrosooxazolidine (17). The two most likely explanations by which nitrosooxazolidines may be formed in metalworking fluids are (1) simple nitrosation of oxazolidine antimicrobials; and (2) nitrosation of primary beta-hydroxy amines (18). The latter reaction is an example of the conversion of a primary amine into a nitrosamine.

It has been demonstrated that beta-hydroxynitrosamines such as NDE1A may undergo retroaldol-type reactions in strongly alkaline medium to form nitrosamines of lower molecular weight (19, 20). In some cases the nitrosamines formed from the retroaldol reaction are more potent animal carcinogens than is NDE1A.

Nitrosamine formation in metalworking fluids appears to be affected by several factors. These factors can be divided into the following categories: (1) catalysts; (2) accelerators; (3) inhibitors; and (4) physical properties affecting nitrosamine formation. Although the presence of nitrite in metalworking fluids leads to high concentrations of nitrosamines, nitrite-free metalworking fluids have also been shown to accumulate nitrosamines (19).

Catalysis of Nitrosamine Formation

Formaldehyde, released from certain antimicrobials in metalworking fluids, activates amines toward nitrosation by nitrite (21). This reaction enhances nitrosation in neutral and basic medium. Since metalworking fluids are typically of pH 9-11, formaldehyde released from antimicrobials into metalworking fluids may be a factor that could enhance nitrosamine formation in this medium (20).

Thiocyanate, an anion normally secreted in human saliva, also catalyzes the nitrosation of amines by nitrite (22). The mechanism of the reaction is thought to proceed through the formation of nitrosylthiocyanate and subsequent reaction with amine to form the nitrosamine. This reaction, originally investigated to assess the potential of nitrosamine formation in the human digestive system, can be related to metalworking fluids since saliva has been found in such fluids (23, 24). Furthermore, the levels of thiocyanate in the saliva of non-smokers contains about 50 mg/l while in smokers the saliva contains 3 to 4 times this concentration (25). Fortunately, thiocyanate exerts its strongest effect at pH 1.5 and becomes less effective as the pH increases. At the relatively high pH of metalworking fluids the catalytic effect of thiocyanate (from saliva) on nitrosamine formation may thus be somewhat lessened.

Microorganisms have been shown to catalyze the formation of nitrosamines from secondary amines in the presence of nitrite (26). The amount of nitrosamine formed, however, increased as the basicity of the parent amine decreased, presumably due to the increase in the amount of unprotonated amine present (27). This reaction is especially important with respect to metalworking fluids since water-based fluids are inevitably contaminated by microbes and fungi. Microbes are thought to catalyze nitrosamine formation by lowering the pH of the medium or catalysis by one or more unidentified metabolic products.

Accelerators of Nitrosamine Formation

Chemical complexes of various transition metals have been shown to promote N-nitrosation (28). These metal complexes include ferrocyanide, ferricyanide, cupric ion, molybate ion, cobalt species, and mercuric acetate. All of the reactions are thought to proceed by oxidation-reduction mechanisms. However, such promotion may not be characteristic of transition metal complexes in general, since ferricyanide ion has been shown to promote nitrosation in metalworking fluids, whereas ferric EDTA does not (20). Since the metalworking operation generates metal chips and fines which build up in the fluids, this reaction could be of significance in the promotion of nitrosamine formation in water-based metalworking fluids.

Biocides that function as formaldehyde-releasers comprise about 60% of total sales of antimicrobials (29). Thus, such antimicrobials are expected to be common additions to metalworking fluids. Examples of formaldehyde-releasing antimicrobials are tris(hydroxymethyl)nitromethane, trivially called tris nitro, 4,4'-(2-ethyl-2-nitromethylene)dimorpholine, and 4-(2-nitrobutyl) morpholine. Experiments involving the formaldehyde-releaser 1,3,5-trimethylhexahydro-s-triazine have shown that this antimicrobial exerts a significant catalytic effect on nitrosamine formation in metalworking fluids (20). In fluids containing diethanolamine the nitrosamine yield reached 50 ppm in 2 days at room temperature. An antimicrobial structurally similar to tris nitro is bronopol, which is widely used in cosmetics and shampoos. A feature common to these compounds is the C-nitro group, which is also thought to release nitrite (30). In the presence of amines, C-nitro compounds are indeed nitrosating agents, and their nitrosation potential increases with addition of electron-withdrawing groups (31). As expected, it has been shown that the rate of formation of NDElA from bronopol and triethanolamine decreases as the pH is decreased from 6 to 4 (32). Thus C-nitro-containing, formaldehyde-releasing antimicrobials possess in the same molecule both the nitrosating agent and the catalyst for its reaction with amines.

Surfactants that form micelles have also been shown to accelerate the formation of nitrosamines from amines and nitrite (33.) A rate enhancement of up to 800-fold was observed for the nitrosation of dihexylamine by nitrite in the presence of the cationic surfactant decyltrimethylammonium bromide (DTAB) at pH 3.5. A critical micelle concentration (CMC) of 0.08% of DTAB was required to cause this effect, which was attributed to a micelle with the hydrocarbon chains buried in the interior of the micelle. The positively-charged ends of the micelle would then cause an aggregation of free nitrosatable amine relative to protonated amine and thus lead to rate enhancements. Since surfactants are commonly used in water-based fluids (25-50% lubricating agent or 10-20% emulsifier in concentrates), concentrations above the CMC of a micelle-forming surfactant could enhance the formation of nitrosamines.

The massive contamination of NDElA in alkaline synthetic fluids (3%) found by Fan et al (7) cannot be explained by known nitrosation kinetics of di- or triethanolamine. Instead, more powerful nitrosation routes, possibly involving nitrogen oxide (NO_x) derivatives (e.g., NO_2, N_2O_3) may be responsible for the amounts of NDElA in these products (34). In fact, a nitrite-free commercial concentrate was shown to accumulate NDElA up to about 100 days at which time the levels dropped dramatically (19). Inhibition of NO_x contaminants may be an effective route to the inhibition of nitrosamine formation in metalworking fluids.

Inhibition of Nitrosamine Formation

Inhibitors of nitrosation generally function by competing with the amine for the nitrosating agent. An inhibitor would thus react with nitrite at a faster rate than with amines. The inhibition reaction has recently been reviewed (35). The ability of ascorbate to act as a potent inhibitor of nitrosamine formation has resulted in the use of the vitamin in nitrite-preserved foods and pharmaceuticals. Furthermore, the effectiveness of ascorbate in inhibiting nitrosamine formation is dependent on (1) the concentration of ascorbate (an excess is required); (2) pH (ascorbate is nitrosated 240 times more rapidly than ascorbic acid); (3) the reactivity of the amine toward nitrosation; and (4) the extent of oxygenation of the system.

In addition to ascorbate and its derivatives, other substances including alpha tocopherol, glutathione, urea, ammonium sulfamate, an unidentified component of milk, sodium sulfite, and cysteine have been shown to inhibit nitrosation by competing for the available nitrosating agent.

Phenols are rather common antimicrobial components of metalworking fluids; however, their use in recent years has been declining (36). The inhibition of nitrosation by phenols has recently been reviewed (35). In general, phenolic compounds inhibit nitrosation by reacting with nitrite (phenol reacts with nitrite 10,000 times faster than with dimethylamine), but the intermediate nitrosophenol is unstable and enhances nitrosation. "The overall effect is dependent on the steady state concentration of the nitrosophenol and the relative degrees of retardation and enhancement exerted by the phenol and the nitrosophenol, respectively (35)".

An extremely desirable method to completely eliminate nitrosamine formation in metalworking fluids would entail replacing any nitrosatable amines with non-nitrosatable amines. Primary amines can be converted to stable nitrosamines only via deaminative self-alkylation, normally a low yield process (18). Glycolamine, a dimer of monoethanolamine, exhibited promising results in a preliminary study (20). In this study, NDElA and nitrosomorpholine were each produced in 10 ppm yield only when the glycolamine control fluid was held at 100°C for 48h. Furthermore, highly substituted secondary amines have been suggested as "safe amine" substitutes for nitrosamine precursors in other products (37).

Physical Properties Affecting Nitrosamine Formation

The effect of temperature on nitrosation has not been studied in great detail. Since nitrosation in aqueous solution is reversible, and the nitrogen-nitrogen partial double bond is heat labile [the dissociation energy of the N-N bond in dimethylnitrosamine is at least 30 kcal/mole (38), it would be expected that at a particular temperature, a steady state concentration would be reached, after which time the nitrosamine would begin to decompose. A factor that could complicate the understanding of this process is that heating may change the mechanism of the nitrosation-denitrosation reaction (e.g., from an ionic to a free radical reaction). Since metalworking fluids experience extremely high temperatures at the point of contact between the machine and the metal, and because the average temperature of the fluid in the basin during use is about 40°C (105°F), temperature may play an important role in nitrosamine formation in metalworking fluids. Preliminary unpublished results indicate that a nitrite-containing fluid that was heated experienced a 5000-fold increase in NDElA compared to the unheated control (20).

Nitrosamine levels in concentrates containing nitrite have been shown to increase during storage (39). Since a concentrate could contain up to 45% triethanolamine and 18% nitrite (7), the concentrate provides an ideal situation in which reaction may occur. In this study, conducted over a 5 to 7 month period, the concentration of NDElA increased from 400 to 800 ppm.

The pH of a metalworking fluid must be kept above neutrality in order to prevent acid corrosion of the metal. In vitro, acid catalyzed nitrosation is optimized at pH 3.5 (40); however, it has been shown that in the presence of other catalysts, aqueous solutions of amines and nitrite leads to significant yields of nitrosamines at room temperature over the pH range of 6.4 to 11.0 (41). Furthermore, C–nitro–containing, formaldehyde–releasing biocides, such as bronopol or tris nitro, exert their potential catalytic effect in alkaline solution. It would thus be desirable to determine the optimum pH for a metalworking fluid that would lead to the lowest concentration of nitrosamine possible.

In summary, it appears that TSCA may have had several effects on the metalworking fluids industry. The reporting requirements of Section 8 have demonstrated an awareness to both EPA and industry, not only of the presence of nitrosamines in metalworking fluids, but of their intrinsic mechanism of action. The TSCA program may have stimulated research in areas of replacing nitrite and may have increased the awareness of the various factors involved in the formation of nitrosamines in metalworking fluids. By enabling EPA and industry to evaluate and/or study the various aspects of metalworking fluids on nitrosamine formation, the TSCA program has encouraged nitrosamine-free nitrite substitutes to be introduced into commerce.

A review of the literature has revealed that several factors associated with metalworking fluids may enhance or control the formation of nitrosamines in metalworking fluids. If nitrite is present in the concentrate with appropriate amines the nitrosamine levels can reach the part–per–hundred level. Yet even nitrite–free metalworking fluid concentrates have been shown to contain part–per–million quantities of nitrosamines.

Factors that catalyze or promote nitrosation are metal complexes, formaldehyde (and other carbonyl compounds), thiocyanate, microorganisms, certain antimicrobials, micelle–forming surfactants, nitrogen oxides, and storage of concentrates (time). Various compounds exist that inhibit nitrosation, and some may be compatible with metalworking fluids (ascorbate, sulfite). The normally alkaline pH of metalworking fluids inhibits acid catalyzed nitrosation, and temperature appears to affect nitrosamine levels as well, albeit at this time the effect is not well understood. The use of certain non–nitrosatable amines and agents that react with nitrogen oxides probably would provide the greatest reduction in, if not elimination of, nitrosation in metalworking fluids.

Thus, with increased awareness and careful consideration of the factors described in this paper, it may be possible to design and control a nitrosamine–free metalworking fluid.

Acknowledgments

The author would like to express appreciation to Dr. Roger L. Garrett, Dr. Paul H. Bickart, Dr. Larry K. Keefer, and Dr. Joseph E. Saavedra for invaluable discussions.

Literature Cited

1. Holmes, P. M. Tribology International 1977, 47-55 (Feb.).
2. Gosselin, R. E., Hodge, H. C., Smith, R. P., Gleason, M. N. "General Formulations." In Clinical Toxicology of Commercial Products; 4th Ed. Baltimore: The Wilkins Co., 1976, p. 158.
3. Bennett, E. O. Lub. Eng. 1979, 3, 137-144.
4. Cookson, J. O. Tribol. Internat. 1977, 29-31 (Feb.).
5. Morton, I. S. Ind. Lub. Tribology 1971, 57-62 (Feb.).
6. Holmes, P. M. Ind. Lub. Tribology 1971, 47-55 (Feb.).
7. Fan, TY., Morrison, J., Rounbehler, D., Ross, R., Fine, D., Miles, W., and Sen, N. Science 1977, 196, 70.
8. Druckrey, H., Preussman, R., Ivankovich, S., and Schmahl, D. Z. Krebsforsch. 1967, 69, 103-201.
9. Hilfrich, J., Schmeltz, L., and Hoffman, D. Cancer Letters 1978, 4, 55.
10. Lijinsky, W., Reuber, M. D., and Manning, W. B. Nature 1980, 288, 589-590.
11. Wishnok, J. S. "N-Nitrosamines"; In Kirk-Othmer Encyclopedia of Chemical Technology. Vol. 15. M. Grayson and D. Eckroth, eds.; John Wiley and Sons: New York, 1981, pp. 988-996.
12. Gustavsen, A. J. "Nitrite Replacement"; Society of Manufacturing Engineers Clinic - Modern Metalworking Fluids. 1981, September 29-October 1.
13. Zingmark, P. A., and Rappe, C. Ambio 1976, 5, 80-81.
14. Natl. Inst. Occupational Safety and Health. Current Intelligence Bulletin: "Nitrosamines in cutting fluids"; Rockville, MD: U.S. Dept. Health, Education, and Welfare. DHEW Pub. NIOSH 78-127.
15. Chemical Week "Top of the News: Warning on Oils"; Oct. 20 1976, p. 20.
16. Chem Eng News "Concentrates" Oct. 18, 1976, p. 12.
17. Stephany, R. W., Freudenthal, J., and Schuller, P. L., Rec. Trav. Chim. 1978, 97, 177.
18. Saavedra, J. E. J. Org. Chem. 1981, 46, 2610-2614.
19. Loeppky, R. N. ACS Abstr. 1980, New Orleans, ORGN 418.
20. Loeppky, R. N., Hansen, T. J., and Keefer, L. K. Fd. Cosmet. Toxicol. ("Reducing Nitrosamine Contamination in Cutting Fluids"), 1982, in press.
21. Keefer, L. K., and Roller, P. P. Science 1973, 181, 1245.
22. Bolyand, E., and Walker, S. A. Arzneim.-Forsch. 1974, 24, 1181-1184.

23. Smith, T. H. F. Lub. Eng. 1969, 313-320.
24. Nehls, B.L. J. Am. Soc. Lub. Eng. 1977, 179-183.
25. Densen, P. M., Davidow, B., Bass, H. E., and Jones, E. Archs. Envir. Hlth. 1967, 14, 865.
26. Yang, H. S., Okun, J. D., and Archer, M. C. J. Agric. Food Chem. 1977, 25, 1181-1183.
27. Hawksworth, G. M., and Hill, M. J. Br. J. Cancer 1971, 25, 520-526.
28. Keefer, L. K. "Promotion of N-nitrosation reactions by metal complexes." In Environmental N-Nitroso Compounds. Analysis and Formation. E.A. Walker, P. Bogovski, and L. Griciute (Eds.). IARC Scientific Publication No. 14. Lyon, France. 1976, pp. 153-159.
29. Rossmoore, H. W. J. Occupational Med. 1981, 23, 247-254.
30. Schmeltz, I., and Wenger, A. Fd. Cosmet. Toxicol. 1979 17, 105-109.
31. Fan, T-Y., Vita, R., and Fine, D. H. Toxicol. Lett. 1978, 2, 5-10.
32. Ong. J. T. H., and Rutherford, B. S. J. Soc. Cosmet. Chem. 1980, 31, 153-159.
33. Okun, J. D., and Archer, M. C. J. Natl. Cancer Inst. 1977 58, 409-411.
34. Fine, D.H. "N-Nitroso compounds in the environment." In Advances in Environmental Science and Technology. Vol. 10., J. Pitts and R. Metcalf (Eds). J. Wiley and Sons., New York. 1980, pp-39-123.
35. Woo, Y-T., and Arcos, J.C. "Environmental Chemicals." In Carcinogens in Industry and the Environment. J.M. Sontag (Ed.). Marcel Dekker, New York. 1981, pp. 168-281.
36. Holtzman, M. "Microbiology of metalworking fluids." Society of Manufacturing Engineers Clinic--Modern Metalworking Fluids. 1980, September 29-October 1.
37. Preussmann, R., Spiegelhalder, B., and Eisenbrand, G. "Reduction of Human Exposure to Environmental N-nitroso compounds. In N-Nitroso Compounds. R. A. Scanlan and S. R. Tannenbaum, eds. ACS Symposium Series No. 174. American Chemical Society, Washington, D. C. 1981, p.217.
38. Fridman, A. L., Mukhametshin, F. M., and Novikov, S. Russ. Chem. Rev. 1971, 40, 34-49.
39. Zingmark, P. A., and Rappe, C. Ambio 1977, 6, 237-238.
40. Mirvish, S. S. J. Natl. Cancer Inst. 1970, 44, 633-639.
41. Smith, R. V. Chemical Times and Trends 1980, 35-42 Jan.

RECEIVED December 28, 1982

Impact on Public Health

MICHAEL J. LIPSETT

SRI International, Menlo Park, CA 94025

Regulations promulgated under the Toxic Substances
Control Act (TSCA) may have a beneficial impact on
public health, though such an impact will be
difficult to measure or to estimate. Such an
evaluation of TSCA's effects is problematic because
of: the difficulty in isolating impacts of TSCA
regulations from other environmental or
occupational health statutes; the insensitivity of
epidemiologic studies in detecting chronic effects
of low-level chemical exposure; and the preventive
nature of the Act, which subjects new as well as
existing chemicals to regulation. A more practical
difficulty exists in that, five years after the
enactment of TSCA, few chemical substances have
been subject to regulation. This paper will
discuss why an evaluation of the health impact of
TSCA must of necessity remain somewhat speculative.

I have been asked to discuss the human health impacts of
TSCA. Any examination of such "impacts" of the Act should focus
on effects that can be measured or estimated. However, in cases
where the statutory goals are primarily preventive in nature,
measurement or even estimation of health benefits may prove
elusive. Although TSCA contains language that appears to
require some consideration of the impact of regulation, it is
unlikely that Congress intended that precise quantitative
evaluation of the effects of TSCA be undertaken. As we shall
see, such an evaluation is not feasible.

Out of the universe of potential health effects that could
be evaluated, I will focus on those specifically designated in
the Act itself--cancer, birth defects and gene mutations. This
is not to say that other potential health impacts of chemical
exposure are unimportant. Rather, these three named effects
represent a relatively circumscribed basis by which to evaluate
the Act in terms of its explicit priorities.

0097-6156/83/0213-0169$06.00/0
© 1983 American Chemical Society

In the first section of my talk I hope to show why measurement of these effects as a function of regulatory actions under TSCA is not practical. If such effects are not measurable, then for regulatory purposes they must be estimated, usually by extrapolation from animal experiments. I will briefly indicate that quantitative extrapolation is an uncertain business. In the second section, I will summarize TSCA's probable impact on health, methodological difficulties in measurement notwithstanding. I will conclude with some remarks about recent regulatory pronouncements which seem to indicate that if past policies have had little discernible health impact, future ones may have even less.

Practical Limitations on Measurement of Health Impact

Assuming, for the sake of argument, that some far-reaching regulations had been promulgated limiting human exposure to one or more substances suspected of causing cancer, birth defects, or gene mutations, it would be difficult, if not impossible, to measure any effect on the incidence of these conditions attributable to such regulations. Some conceptual and practical impediments to such measurement include:

. Difficulty isolating the effect of TSCA from consumer protection and other environmental and occupational health statutes and regulations.

. Relative insensitivity of epidemiologic studies in detecting long-term effects of low-level chemical exposures.

. Inability in most instances to detect potential beneficial health effects from reduced exposure to chemicals due to the chronic nature and multifactorial etiologies of the conditions in question.

. The preventive nature of the Act, under which new as well as existing chemicals may be subject to regulation.

Isolation of TSCA's Effect. There are at least 17 federal statutes and numerous state laws purporting to regulate human exposure to hazardous substances. While many chemical substances that may pose chronic health hazards are excluded from regulation under TSCA (e.g., pesticides, drugs, cosmetics, alcoholic products, food additives, tobacco), others are potentially subject to concurrent regulation under several statutes, including TSCA. For example, asbestos, a well-recognized human carcinogen, may be regulated by EPA under TSCA, and at the same time is subject to regulation by the same agency under the Clean Air and Clean Water Acts. The Consumer

Product Safety Commission regulates the presence of asbestos in various consumer products. Workplace ambient standards for asbestos have been established under the Occupational Safety and Health Act, though the basis for such regulation is not carcinogenicity. Many other chemical substances are subject to overlapping regulation with respect to different aspects of human activity. Thus, with the exception of chemical exposures uniquely subject to regulation under TSCA, the potential influence of this statute on overall human exposure to hazardous chemicals will be diluted by the effects of regulations promulgated under other environmental, occupational, and consumer protection statutes.

Insensitivity of Epidemiologic Studies. There are obvious ethical and legal limitations on administering potential carcinogens, mutagens and teratogens to humans in an experimental setting. Therefore, to evaluate whether particular chemical agents increase the risk of chronic health effects, one must rely on epidemiologic studies of populations exposed to such agents. Since epidemiologic investigations are not controlled experiments, and since they are usually undertaken retrospectively, they are subject to limitations that affect their sensitivity to detect chronic effects. One of the most important limitations is the lack of good exposure data.

With the exception of industrial hygiene data for selected industries and ambient air quality monitoring for Clean Air Act "criteria" pollutants, there is little more than sporadic sampling of environmental media and human environments for chemicals that could be subject to TSCA regulation. In attempting to study whether human exposure to a particular chemical is associated with a given chronic disease outcome, one must try to ascertain past exposure to that chemical. However, in settings other than the workplace, measurements of this kind are virtually nonexistent. In addition, people who work with chemicals are typically exposed to multiple substances, and such overlapping exposures may be difficult to control for either in the design or the analysis of epidemiologic studies. In non-occupational contexts, uncontrolled and unmonitored low-level exposures to multiple substances in food, air, and water are a typical feature of everyday life that may make epidemiologic identification of independent risk factors even more difficult. For most chemicals potentially subject to TSCA regulation, epidemiologic studies will not be able to resolve questions of associations of chemical exposures with particular disease outcomes.

Chronic Nature of Diseases in Question. Cancer is a disease characterized in most cases by a latency period of 15 to 40 years. That is, there is a lag of 15 years or more between initial exposure to a carcinogen and the manifestation of the

disease. (The principal exceptions to this observation are
cancers of the hematopoetic tissues, which have a minimum
latency period of around 5 years.) Thus, to measure the effects
of some hypothetical regulations, one would have to look at
cancer incidence 15 or more years after their adoption among a
defined group of people who would otherwise be exposed to the
chemical or chemicals in question.

For existing chemicals that are recognized human
carcinogens, it might be possible to estimate the number of
cancers avoided by reducing exposure over a lifetime. In theory
this could be calculated using dose-response information to
estimate the benefits of a percentage reduction of exposure to
particular chemical substances. Unfortunately, such human
dose-response data are unavailable for all but a few
carcinogens, and even for these, the effects of low doses can
only be guessed.

With respect to birth defects, the time lag between
exposure and outcome is not so much of a problem as is the case
with cancer. However, attributing human birth defects to
chemical exposures (other than pharmaceutical products, smoking,
alcohol, and a few occupational exposures), is difficult. Even
if a particular substance is capable of causing birth defects in
humans, the occurrence of such an outcome depends on the dose,
on the route of maternal exposure, and on the timing of the
exposure. Nevertheless, epidemiologic investigations
(principally case-control studies) have ascertained causal
relationships between chemical exposures and adverse
reproductive outcomes, such as congenital anomalies. Most fetal
defects are incompatible with fetal survival and result in
spontaneous abortion. This potentially sensitive parameter of
birth defects--i.e., spontaneous abortion--is not routinely
monitored, and therefore provides no baseline from which to
measure the potential effect of reducing a given chemical
exposure.

Defects among live births are, however, routinely but not
systematically monitored nationwide by the Centers for Disease
Control (CDC). This monitoring system probably does not convey
an accurate picture of the prevalence of birth defects, though,
since it is based on hospital discharge abstracts, which often
do not contain information about any but the most severe and
obvious congenital anomalies. These abstracts contain little or
no information on maternal factors, such as occupational
exposures. Thus this system would probably not be sensitive
enough to detect the effects of eliminating or significantly
reducing chemical exposures, if such exposures do in fact have a
major influence on the incidence of birth defects. Recent
estimates of the percentage of birth defects attributable to
environmental exposures of all kinds, including smoking, drugs,
infections, radiation, and general environmental chemicals,
indicate that this category probably represents about 10% of
birth defects.(1) Most congenital anomalies (about 2/3) are of

unknown etiology. Chemicals that have been linked to birth
defects or other reproductive effects have been detected as
etiologic agents because pregnant women have had relatively
large, well-documented exposures in the form of ingestion of
drugs or alcohol, accidental poisoning, or occupational
exposure. Epidemiologic studies are, in general, too
insensitive to detect effects at lower exposure levels, unless
the substance of interest is extremely potent.

With respect to the third chronic health effect mentioned
specifically in TSCA—gene mutations—again the nature of the
effect in question is such that historical measurement is not
feasible. Mutagenic properties are studied in microbial, cell
culture, and animal systems. Human body fluids can be monitored
for the presence of mutagenic substances, but this does not
actually measure genetic effects. While several methods for
monitoring mutational events in humans are being developed, none
is ready yet for general use.(2) There are, however,
cytogenetic techniques to examine potential genetic effects of
chemical exposure on humans. These look at a higher level of
genetic organization—chromosomal and chromatid aberrations.
Some studies of persons exposed to chemicals demonstrate an
increase of such aberrations correlating with the time of
exposure. In other investigations, individual variability has
overshadowed any differences that might be attributable to
chemical exposure. Such investigations are complicated by a
lack of knowledge about the frequency and persistence of
spontaneous chromosomal aberrations. Furthermore, there is
little evidence linking such changes to specific diseases,
though intuitively one would expect such an association.
(Several types of human cancers have been reported to be
associated with specific chromosomal rearrangements.)

In general, mutational events are considered detrimental.
Teleologically speaking, this is why living systems have evolved
multiple DNA repair mechanisms. Some investigators have
estimated that about 90% of known carcinogens act through
mutational mechanisms.(3) Furthermore, many human diseases,
including sickle cell anemia, thalassemia, mucopolysaccharidoses
and others, are known to have a genetic basis. However, my
point is that while the state-of-the-art of genetic toxicology
is rapidly evolving, it is not yet capable of measuring
mutagenic events in complex human systems. Cytogenetic
techniques can detect chromosomal effects, though experience
with these techniques is limited.

Regulation of New As Well As Existing Chemicals. An
evaluation of TSCA's impact would differ from that of most
environmental statutes in that the former purports to regulate
new as well as existing chemicals. The objective of the
premanufacturing notification (PMN) system under section 5 is to
permit EPA to make a reasoned evaluation of new chemicals'

toxicities prior to their production and distribution in
commerce. EPA has authority to impose a broad spectrum of
controls to prevent or minimize human and environmental exposure
to chemicals that could result in disease. This authority has
been used mainly to require more extensive testing of several
suspect chemicals.

In the cases where EPA has formally required such testing,
the manufacturers have withdrawn their applications and
suspended plans to produce the chemicals. While the decision
not to produce a potentially toxic substance may serve the goal
of TSCA to identify and prevent hazards before people are
exposed, how can one quantitate the health impact of the
manufacturers' decisions? I pose this question rhetorically,
since EPA's requests for additional testing stemmed from a data
base inadequate to assess risk. In other words, since there was
not enough information in the first place to know whether there
might even be a substantial health risk, it would be impossible
to estimate the health impact of deciding not to produce such
chemicals.

By now I hope it is clear that measurement of the human
health impact of TSCA, at least with respect to cancer, birth
defects, and genetic mutations, is not currently feasible, for
both practical and theoretical reasons. Thus, any evaluation of
regulations under TSCA in terms of potential health benefits
must be based on predictions from epidemiologic or clinical
data, and from animal and microbial models. Risk assessment
without human toxicity data is unavoidable under section 5
regulatory decisions concerning new chemicals, though for
regulations of existing chemicals, clinical and epidemiologic
evidence may be available. However, in most cases where
exposure to chemical substances (other than drugs or cigarettes)
has been shown to be associated with cancer or birth defects in
humans, accurate exposure data are not available, and therefore
dose-response curves can be only crudely approximated.

In animal experiments exposures can be carefully
controlled, and dose-response curves can be formally estimated.
Extrapolating such information to the human situation is often
done for regulatory purposes. There are several models for
estimating a lifetime cancer risk in humans based on
extrapolation from animal data. These models, however, are
premised on empirically unverified assumptions that limit their
usefulness for quantitative purposes. While quantitative cancer
risk assessment is widely used, it is by no means universally
accepted. Using different models, one can arrive at estimates
of potential cancer incidence in humans that vary by several
orders of magnitude for a given level of exposure. Such
variations make it rather difficult to place confidence
intervals around benefits estimations for regulatory purposes.
Furthermore, low dose risk estimation methods have not been
developed for chronic health effects other than cancer. The

implication of the limitations of risk assessment methodology
and health impact measurement is that, with narrowly defined
exceptions, health benefits of TSCA regulations cannot be
realistically estimated.

Few Effects of TSCA Regulatory Actions

If the health impact of TSCA regulations is not possible to
measure, and if estimates of health benefits of regulations are
difficult to quantify, little has been achieved under TSCA that
could be measured or estimated. In the next section of this
paper, I will be discussing regulation under section 5, which
deals with new chemicals and significant new uses of existing
chemicals, and under section 6, which involves chemical
substances and mixtures known to be hazardous. I will not
discuss actions under section 8, concerning potential health
hazard reporting, nor will I cover the effects of actions
undertaken by the individual states with section 28 grants.

Impacts directly attributable to TSCA regulations include
several proposed and final regulations directed at specific
chemical substances under section 6--polychlorinated biphenyls
(PCBs), dioxin, chlorofluorocarbons (CFCs), and asbestos--and
orders issued under section 5(e).

PCBs. Congress singled out PCBs from all other
environmental contaminants for regulatory attention under
section 6(e) of TSCA. EPA was directed to develop regulations
for labelling, use and disposal, as well as to promulgate rules
for implementing a statutory ban on manufacturing, processing
and distribution of PCBs other than in a "totally enclosed
manner" or in a way that the EPA Administrator considered safe.
In a legal challenge to these regulations early last year, the
D.C. Court of Appeals found EPA's definition of "totally
enclosed uses" to be unsupported by the procedural record, and
directed EPA to rewrite some parts of these regulations.
Subsequently the judge's order was stayed for 18 months to allow
EPA to gather additional evidence. Thus, the reformulation of
these rules will not be completed until later this year or early
1983. Those PCB regulations still on the books may have helped
to diminish human exposure to PCBs, though for the reasons
discussed earlier, the health impact of such diminished exposure
is not measurable.

Dioxin. Two years ago, EPA promulgated a rule prohibiting
Vertac Chemical Corporation from disposing of waste contaminated
by dioxin. Other parties intending to dispose of similarly
contaminated wastes were required to notify EPA 60 days in
advance of their intentions. This order may have prevented some
exposure to this highly toxic substance, though the human health
impact of this single prohibition cannot be calculated.

Asbestos. EPA issued a proposed rule concerning identification and correction of friable asbestos-containing materials in schools. Based on data voluntarily submitted, EPA estimated that at least 8,600 public schools attended by over 3 million children contain such materials. However, EPA reportedly has no information on another 44,000 schools. Classroom concentrations of asbestos fibers in some schools have been found to approximate concentrations in homes of asbestos workers who do not have shower or laundry facilities at work. Since children exposed to asbestos will live long enough to allow the cancer latency period to elapse, the presence of friable asbestos materials in schools represents a potentially enormous public health problem. The final asbestos rule will reportedly be promulgated in the near future. (The rule was published May 27, 1982.) No other regulations regarding asbestos have been issued.

CFCs. All "nonessential" uses of CFCs in aerosol propellents were banned in 1978--the first and only major control action under TSCA not specifically mandated by the statute. This action may have helped to reduce the future incidence of skin cancer by diminishing CFCs' destructive effects on stratospheric ozone. Making appropriate assumptions about rates of ozone depletion and extrapolating from current disease rates, one could estimate a range of cancers avoided because of this prohibition. However, any health benefit due to the ban on aerosol CFC uses may be masked by the continued increase in non-aerosol uses.

All in all, regulatory actions under section 6 are not likely to have achieved a major effect on human health. One reason for this is that under TSCA section 9, regulatory deference is accorded to other statutes and, where appropriate, to other regulatory agencies. Another is that TSCA is conceptually more difficult to administer than other environmental statutes that set target goals and dates for pollution reduction. TSCA focuses instead on the prevention of "unreasonable risks," in which the definition of what is unreasonable depends in part on potential health benefits that are difficult to quantify. While there have been other impediments to the regulation of known hazards, one critical factor has been EPA's de-emphasis of such regulatory actions in favor of gathering data and setting up a system to screen and monitor new chemicals.

PMNs for New Chemicals. How well has the PMN system worked from the perspective of protecting human health? As was noted earlier, there is no way to directly measure the benefits. Nine chemicals have been withdrawn from production as a result of orders requiring more extensive testing. Informal negotiations reportedly resulted in labelling and use restrictions or further testing on about 60 others.(4) There is no way to assess the impact of these actions on health, since the content of these informal negotiations is not public knowledge.

Although TSCA section 2 assigns the responsibility for developing adequate toxicity data to manufacturers and processors of chemicals, it has been staff members of EPA who have been doing most of the toxicologic work under section 5. As of the end of 1980, two-thirds of PMNs submitted contained no toxicity information whatsoever. Preliminary statistics from 1981 indicate that a greater percentage of PMNs during the past year contained more toxicity testing information. Still, there were few chronic toxicity data. The lack of such information has meant that EPA's evaluations have had to be conducted on the basis of structure-activity relationships. Such analyses involve comparing the PMN chemicals to existing ones with similar structures whose toxicities are known.

Structure-activity studies are probably adequate for some substances. An example would be inert polymers whose monomeric components have been well-characterized toxicologically. For other chemicals, analyses based on a review of structural analogues may prove inadequate for at least three reasons. First, minor molecular modifications may have a dramatic effect on toxicologic properties. For instance, 2,6-heptanedione is relatively harmless, while 2,5-heptanedione is a neurotoxin.(5) Second, there may not be any corresponding chemicals for which adequate chronic toxicity data exist, since most existing chemicals have not been subject to such testing. Third, the molecular bases for chronic toxic effects have been thoroughly worked out only for certain classes of mutagens, carcinogens, and antimetabolites. Within these classes, structure-activity analyses can be useful in identifying potential "bad actors," and, indeed, have led to informal requests for more extensive toxicity data or to section 5(e) orders. However, other mechanisms of carcinogenicity and mutagenicity, as well as molecular explanations for teratogenicity, other adverse reproductive effects, neurotoxicity and other chronic toxic effects, have not been well-characterized and cannot be incorporated into structure-activity reviews. Thus, in the absence of testing of new chemicals for chronic toxic effects, the PMN review process probably cannot provide an adequate screening at the present time.

A former Assistant Administrator for Toxic Substances observed that such analyses are "based upon a fundamental lack of information and data. This in turn means that our information will be highly uncertain."(6) On the other hand, prior to the establishment of the PMN system, those chemicals for which EPA requested better data might otherwise have been produced or distributed in commerce with little or no testing whatever.

While there has been little regulatory action under
sections 5 and 6, TSCA may have had some indirect health
effects. For example, chemical companies are more aware now
than they were five and a half years ago, when TSCA was enacted,
about chronic health hazards in the workplace. In general,
therefore, workplace exposures to chronic toxic hazards are
likely to be lower than in the past. To some indeterminate
extent, the compiling of the TSCA inventory and the TSCA
reporting requirements may have played a role in creating a new
awareness of chemical hazards. Other social and legal
developments, however, have probably been more important in the
creation of such an awareness. Among such other influences
would have to be included product liability litigation, OSHA
regulations, union pressures, and a more general increasing
consciousness of potential adverse effects of chemical
production and use, due to greater media coverage of these and
related issues.

Conclusion

In this paper I have tried to show that measurement of
health benefits attributable to TSCA is not feasible. I hope
that in doing so I have not belabored the obvious. For new
chemicals and for most existing chemicals, prospective
evaluation of health benefits to be achieved by various exposure
controls will have to be based on extrapolation from microbial
and animal data. However, while such extrapolation may be
useful in a qualitative sense, quantitative risk assessment
techniques involve considerable uncertainty, and in any case
have not been developed for chronic effects other than cancer.
Measurement or estimation of health impacts under TSCA
would be premature, since relatively little has been done to
regulate new or existing chemicals that could result in health
benefits. The principal exception to this generalization is the
ban on aerosol uses of CFCs, whose chronic effects on human
health derive from their environmental impact rather than direct
biological toxicity. Compared with other environmental laws,
such as the Clean Air Act, the regulatory accomplishments of
TSCA are somewhat insubstantial.
A large part of the difficulty in developing regulatory
initiatives under TSCA may be the lack of specific statutory
direction. Preventing unreasonable risks is harder to implement
as a policy than achieving percentage reductions in air
emissions of particular pollutants. The implementation of this
Act has therefore tended to focus on information-gathering
objectives rather than control activities.
Deciding what are reasonable or unreasonable risks depends
on the quantity and quality of information about costs, risks
and benefits of different levels of production and exposure
controls with respect to a particular chemical or class of

chemicals. To the extent that human epidemiological data are available, regulatory decisions should take them into account. However, to postpone such decisions (as has recently been done in the case of formaldehyde) on the grounds that ongoing or future epidemiologic studies will resolve critical health issues is, in my opinion, misguided.

Such studies (particularly cohort mortality studies) typically take several years to complete, and may not yield definitive answers. Because of the inherent limitations of such investigations, the usual standard of proof of causation in epidemiology is consistent results from multiple studies conducted under different conditions. Such studies cannot disprove the carcinogenicity of a chemical--at best, they can indicate only an upper limit of risk. The expense and relative insensitivity of epidemiologic investigations insure that they will be of limited importance in identifying chronic health effects attributable to specific chemical exposures. Finally, a policy of delay pending the results of epidemiologic studies implies that an apparently higher threshold of certainty regarding health risks must be reached before initiating regulatory action. This would make sections 5 and 6 even more difficult to implement, and would portend that the health impact of TSCA will continue to be of marginal significance.

Literature Cited

1.   Wilson, J. G. "Environment and Birth Defects;" Academic Press: New York, 1973, cited in Klingberg, M. A.; Papier, C. M. "Environmental Teratogens;" in "Contributions to Epidemiology and Biostatistics;" Klingberg, M. A., Weatherall, J.A.C., Eds., Karger: Basel, 1979.
2.   Bloom, A. D., Ed. "Guidelines for Studies of Human Populations Exposed to Reproductive Hazards;" March of Dimes Birth Defects Foundation: New York, 1981, passim.
3.   McCann, J.; Ames, B. "The Salmonella/Microsome Mutagenicity Test: Predictive Value for Animal Carcinogenicity;" in "Origins of Human Cancer;" Hiatt, H. H., Watson, J. D., Winsten, J. A., Eds., Cold Springs Harbor Laboratory: Cold Springs Harbor, New York, 1977, pp. 1431-50.
4.   U.S. Environmental Protection Agency, Office of Toxic Substances. "Priorities for OTS Operation;" Washington, D.C., January 1982.
5.   Schaumburg, H. H.; Arezzo, J. C.; Markowitz, L.; Spencer, P. S. "Neurotoxicity Assessment at Chemical Disposal Sites;" in "Assessment of Health Effects at Chemical Disposal Sites;" Proceedings of a Symposium held in New York City on June 1-2, 1981, Lowrance, W. W., Ed., The Rockefeller University.

6.   Jellinek, S. D. Paper presented at Fuji Techno Systems
      Seminar on the Impact of Regulatory Requirements on
      Chemical Substances, Tokyo, Japan, Oct. 30, 1980, cited in
      U.S. Office of Technology Assessment, "Assessment of
      Technologies for Determining Cancer Risks from the
      Environment;" 1981, p. 148.

RECEIVED November 22, 1982

Quantitative Analysis as a Basis for Decisions Under TSCA

D. WARNER NORTH[1]

Decision Focus Inc., Palo Alto, CA 94304

The language and the legislative history of TSCA leave ambiguous the extent to which formal analytical methods should be used to determine whether a chemical substance or mixture presents or may present an "unreasonable risk" of harm to human health or the environment. Decision makers implementing TSCA confront large uncertainties and great complexity in assessing the available information on chemical toxicity and exposure. Use of a probabilistic methodology such as decision analysis allows uncertainty to be included explicitly in the basis for decision. Case studies on specific chemicals indicate that quantitative approaches based on decision analysis offer significant potential for improvement of the regulatory decision process under TSCA. However, it is important that the analysis be perceived as a framework for discussion, debate, and investigation of sensitive assumptions rather than as a mechanistic formula for determining regulatory decisions.

Passage of the Toxic Substance Control Act (TSCA) in 1976 was widely regarded at the time as a welcome improvement in environmental legislation. Unlike the language of the Clean Air Act or the Delaney Amendment, TSCA avoids calling for absolute elimination of health risks, requiring instead a balancing between the adverse effects on health and the environment and the benefits of a chemical substance or mixture. The impact of TSCA to date has been somewhat disappointing. Environmentalists note that few regulatory decisions have been made under TSCA, and they fear that the Reagan Administration's call for cost-benefit analysis under Executive Order 12291 may result in aggravating EPA's preexisting

[1] Current address: 4984 El Camino Real, Suite 200, Los Altos, CA 94022.

0097-6156/83/0213-0181$06.00/0

tendency toward "paralysis by analysis." Industry, on the other
hand, is troubled by a lack of clear rules for determining what
chemical uses are acceptable and what information should be
gathered on chemical toxicity.

Difficulties with TSCA

TSCA represents a step away from legislative requirements to
eliminate risk and a step toward balancing the beneficial and
adverse consequences of chemical use, but the Act suffers from
ambiguity compared to earlier legislation. The term "unreasonable
risk" is used repeatedly in virtually all the key sections of the
Act but is not defined clearly either in the Act itself or in its
legislative history. Perhaps the closest approach to a clarifica-
tion occurs in the following passage from the legislative history
(1):

> In general, a determination that a risk associated
> with a chemical substance or mixture is unreasonable
> involves balancing the probability that harm will
> occur and the magnitude and severity of that harm
> against the effect of proposed regulatory action on
> the availability to society of the benefits of the
> substance or mixture, taking into account the avail-
> ability of substitutes for the substance or mixture
> which do not require regulation, and other adverse
> effects which such proposed action may have on
> society.
>
> The balancing process described above does not
> require a formal benefit-cost analysis under which a
> monetary value is assigned to the risks associated
> with a substance and to the cost to society of
> proposed regulatory action on the availability of
> such benefits. Because a monetary value often cannot
> be assigned to a benefit or cost, such an analysis
> would not be very useful.

The first part of the passage calls for a balancing of the
probability of harm and the magnitude and severity of that harm
against benefits that might be lost by placing regulations on the
chemical substance or mixture in question. The ambiguity of the
passage is that the balancing process needed for the determination
of unreasonable risk is not described. Rather, the second part of
the passage is phrased negatively: The balancing process does not
require a formal cost-benefit analysis in which monetary value is
assigned to the cost and to the risk.
Given this guidance, how is EPA to reach decisions, and how
can industry understand EPA's decision process so it can
anticipate these decisions in planning its business activities?

Economists, analysts, and spokesmen for the current Administration might assert that despite the caveat in the legislative history that formal cost-benefit analysis is not required, some sort of cost-benefit balancing ought to be used in the decision process. But the application of traditional cost-benefit methods to assessing the consequences of chemical regulation does not account for uncertainty. Although for a large number of chemical substances and mixtures there are grounds to suspect potential adverse effects on human health or ecological systems, it is rarely the case that the extent of these adverse effects can be estimated with any precision. It is also often difficult to foresee the economic consequences of a regulatory decision that will require a chemical to be removed from a major use or manufactured using an untried modification in the production process. Costs and benefits of regulation under TSCA will usually be uncertain. Yet the balancing process must be carried out, explicitly or implicitly, to reach decisions, both within EPA and within individual chemical companies.

If analysis is to be useful in assisting this balancing process, it must deal with uncertainty. How this can be accomplished involves a simple concept: the use of probability as a way of communicating judgment about uncertainty. This concept is common sense to many people. We often communicate using this concept about sporting events (e.g., the probability that the San Francisco Forty Niners will win next Sunday's football game), outcomes of elections (e.g., the probability that the Republicans will maintain control of the Senate in the next election), and weather (e.g., the probability of rain). In these situations the probability numbers serve as summaries of judgment about a multitude of complex factors. The judgments of political and sports experts and weather forecasters may be good or they may be poor. What probability provides is a way to describe uncertainties quantitatively so that we can discuss these uncertainties more precisely and incorporate them into our decision making.

Decision analysis provides a formal theory for choosing among alternatives whose consequences are uncertain. The key idea in decision analysis is the use of judgmental probability as a general way to quantify uncertainty. Decision analysis has been widely taught and practiced in the business community for more than a decade (2-4). It provides a natural way to extend cost-benefit analysis to include uncertainty.

This paper will summarize briefly some work my colleagues and I at Decision Focus Incorporated have carried out for EPA to show how decision analysis might be used to assist decision making under TSCA (5). I will first briefly review the concepts of quantitative risk assessment and cost-benefit analysis to show how decision analysis fits with these concepts and provides a natural way of extending them. Then I will illustrate the approach using a case study on a specific chemical, perchloroethylene.

An Overview of Quantitative Methods for Assessment and Evaluation
of Chemical Risks

     The literature on analysis applied to assessment and evalua-
tion of chemical risks is very extensive. There is wide agreement
that quantitative analysis is useful as a framework for organizing
information, for facilitating communication among the concerned
parties, and for maintaining a separation between scientific
information and the value judgments that are needed to provide a
basis for decisions, but about which people may disagree. There
is also wide agreement in the literature that quantitative
analysis should not be expected to provide a mechanistic process
or formula for selecting regulatory decisions. The responsibility
for these decisions should remain with agency or company management. The difficult tasks of making the value tradeoffs between
harm to health and economic impacts should come from top management, not from analysts or even worse, from assumptions buried in
a computer program. Besides uncertainty and the difficulties of
making tradeoffs between health and economic values, there are
difficulties in dealing with distributional impacts: Who will
receive benefits under a given policy to control a potentially
toxic chemical, and who will receive the costs? Quite often, the
benefits and costs occur to different groups and, sometimes, they
occur at different times. How benefits or costs occurring far
into the future are to be treated is another area of difficulty.
     There are relatively few case studies of quantitative risk
analysis and cost-benefit analysis applied to chemicals posing a
risk to human health and the environment. Those available include
a variety of reports from committees of the National Academy of
Sciences/National Research Council that have employed quantitative
analysis methods or discussed the use of such methods in decision
making by EPA or similar regulatory agencies (6-11). While much
of the analysis is commendable, there is little consistency in
assumptions, methods, or even terminology. Within EPA's Office of
Toxic Substances, it is difficult to identify and characterize
quantitative analysis that might be used to facilitate the
balancing process described in the TSCA legislative history.
     There is broad agreement on what questions need to be
addressed in carrying out quantitative analysis, but some disagreement on what to call the various steps of the process. The
terminology I shall use below is somewhat arbitrary; my main
purpose is to review the concepts.

Hazard Identification. Does a chemical substance or mixture cause
adverse human health effects, such as cancer, birth defects,
neurological damage, etc? While it would be useful to have an
unequivocal positive or negative answer to this question, that is
rarely possible. The usual situation is that similarity to
chemicals known to be toxic, toxicological testing in cellular
systems or whole animals, and/or epidemiological studies provide

evidence for suspecting that a given chemical agent may cause adverse health effects.

Unit Risk Assessment/Assessment of Dose Response Relationship. Given that a chemical agent can induce cancer or some other adverse health effect in humans, what is the incidence of the effect for a given level of exposure or dose? This question can rarely be answered very precisely because for most chemical agents, human data are not available, and even when such data are available, it is usually very difficult to establish the doses of toxic chemicals to which people were exposed in an epidemiological study. The usual situation is that dose response relationships are estimated from animal bioassay data. EPA's Carcinogen Assessment Group (CAG) routinely produces such unit risk estimates for suspected carcinogens, using a standard set of statistical procedures and assumptions (12,13).

Exposure Assessment. What is the dose or the level of exposure of humans to the chemical agent? This question must be asked in the context of a given policy for controlling the uses and dissemination into the environment of a chemical agent. This control policy might be the present situation, a possible new regulatory policy, or a policy that a chemical manufacturer or distributor could choose to impose on his product. It is usually appropriate to assess the exposure of specific groups of people, which may depend on occupation, life style, purchases and uses of certain products, etc.

Risk Assessment. What is the incidence of the adverse health effects from the chemical agent? This crucial question for regulatory decision making might be answered by combining the unit risk assessment with the exposure assessment. As in the exposure assessment, the question must be addressed in the context of one or more specific control policies.

Risk Evaluation/Cost-Benefit Assessment. Given that risk assessment gives a means of estimating the change in the incidence of adverse health effects that will result from shifting to a new control policy, how are these health impacts to be balanced against the economic and other consequences that the policy change will have? This question involves making value judgments about health consequences, about economic consequences, and about the tradeoffs between them. These value judgments are very sensitive. There is great concern that using monetary values may be misleading, inappropriate, or unethical, yet there is little disagreement that the necessity of making decisions requires tradeoffs between health and economic consequences to be made, either explicitly or implicitly, in the decision process. Despite its difficulties, cost-benefit analysis is becoming increasingly important in regulatory decision making. Executive Order 12291 now requires that federal agencies carry out an assessment of costs and benefits for proposed major regulations. The assessment of the incidence of adverse health effects in risk assessment plus the evaluation or balancing of health and other consequences of control policies in risk evaluation can provide an explicit basis for decision making. If the risks posed by a chemical agent are judged to be unacceptable under the current control policy, a set of possible new control policies is developed, and the best of these is selected. Unfortunately, judgments about the acceptability of risks are apt to be highly controversial, and often there is bitter disagreement about the choice of the "best" regulatory policy. Much of this controversy comes from the methods that are used to deal with uncertainty in the risk assessment process. Conservative, worst-case assumptions are often used in agency risk assessments where precise predictions cannot be made from the available scientific data. For example, tumor response data from the most sensitive animal species at very high dose levels may be extrapolated using a linear nonthreshold model to assess the extent of human cancer that will result from low dose exposure. The two assumptions that the dose response relationship can be extrapolated from the most sensitive animal species to humans and from high doses to low doses using a linear model are each a mixture of scientific judgment and value judgment that it is better to overestimate human health impacts than to underestimate them.

Decision Analysis. An alternative to making assumptions that select single estimates and suppress uncertainties is to use decision analysis methods, which make the uncertainties explicit in risk assessment and risk evaluation. Judgmental probabilities can be used to characterize uncertainties in the dose response relationship, the extent of human exposure, and the economic costs associated with control policies. Decision analysis provides a conceptual framework to separate the questions of information, what will happen as a consequence of control policy choice, from value judgments on how much conservatism is appropriate in decisions involving human health.

A Case Study Application: Perchloroethylene

I now shall present a summary of an application of decision analysis to a specific chemical, perchloroethylene (PCE), a widely used dry cleaning solvent (also called tetrachloroethylene). Full details of this application are presented in an EPA report (5). Perchloroethylene was selected for us by the staff of the EPA Office of Toxic Substances as representative of chemicals on which EPA needed to make an unreasonable risk determination under TSCA. Our analysis was carried out as an exercise in methodology development and not to support any specific regulatory activities by EPA concerning perchloroethylene.

Health Effects of Perchloroethylene. The basis for concern about perchloroethylene was primarily the result of an NCI bioassay, indicating that PCE induced hepatocellular carcinomas in B6C3F1 mice. A similar NCI bioassay on rats had given a negative result for PCE, as had a rat bioassay carried out by Dow Chemical. EPA's Carcinogen Assessment Group (CAG) had prepared a risk assessment based on the bioassay data (14). A meeting of the EPA Science Advisory Board Subcommittee on Airborne Carcinogens was called to review the CAG assessment as part of the determination of whether PCE should be regulated as an airborne carcinogen (15).

The CAG risk assessment included a unit risk estimate of the dose response relationship made following CAG's standard procedures (12). A review of the SAB transcript showed that alternative assumptions were viewed as plausible by the members of the Airborne Carcinogens Subcommittee. CAG had fitted its usual multistage model to the data using a 95% upper confidence limit, a procedure which leads to linear low dose extrapolation (16). Yet the SAB scientists noted evidence that PCE does not act directly on DNA, but indirectly through cellular toxicity. Given an epigenetic mechanism, a nonlinear dose response relation might plausibly be expected. Similarly, while CAG has used the B6C3F1 mouse data as the basis for its extrapolation, scientists at the SAB meeting argued that the rat was more representative of the human metabolism. Finally, while CAG had extrapolated dose level from animal to human using relative surface area, many scientific groups have recommended daily dose per unit of body weight as an appropriate scaling procedure.

Three instances were thus identified where there was uncertainty whether CAG's assumption was right, or whether there might be an alternative assumption that was more appropriate. The sets of assumptions are summarized in Table I. If one assumes for simplicity that for each of the three issues either the CAG assumption or the alternative assumption is correct, then we have eight possible combinations or cases, only one of which represents the correct dose response relationship. Using the methods of decision analysis, we might assign judgmental probabilities to the eight cases. Such probabilities were used in our report, although the numbers are strictly illustrative. The probability numbers are less important than the concept of using a variety of cases based on alternative plausible assumptions. As we shall describe below, the magnitude of the change in estimated cancer incidence from these changes in the dose response assumptions is nearly five orders of magnitude.

Exposure Assessment. Since perchloroethylene is used as a dry cleaning solvent and PCE vapor is easily monitored, estimates of PCE exposure are relatively straightforward to make from existing data in the literature. Table II summarizes the results. Based on NIOSH data (17), machine operators are exposed to an average of about 30 ppm of PCE vapor during the working day, equivalent to a

Table I. Dose Response Assumptions for PCE Case Study

Issue	CAG Assumption	Alternative Assumption
Choice of species	B6C3F1 mouse most sensitive species	Rat better represents human metabolism
Scaling of dose from animal to human	Ratio of surface area	Ratio of body weight
Low dose extrapolation	Multistage model (equivalent to linear with use of upper confidence limit)	Nonlinear response because of epigenetic mechanism (quadratic relation used as representative)

(Eight Combinations of Assumptions Possible)

Table II. Exposure Estimates for PCE Vapor

Classes of People Exposed	Number Exposed	Annual Average Exposure $(\mu g/m^3)$
Workers		
Machine Operators	17,000	45,000
Other Workers	130,000	10,000
Workers in Coin-Operated Facilities	33,000	6,000
Service Users		
Commercial Customers	50 million	5
Coin-Op Cleaners	25 million	10
Coin-Op Laundry	37 million	38
Urban Residents	95 million	0.2-4

continuous exposure of 45,000 $\mu g/m^3$. Other workers in commercial and industrial dry cleaners are exposed at a lower level, 10,000 $\mu g/m^3$, and workers in coin operated laundromat-dry cleaning facilities have an estimated exposure level of 6,000 $\mu g/m^3$. The number of workers exposed is based on projections from industry and census data.

Users of dry cleaning services are exposed when they visit the dry cleaning facilities and a lesser extent from cleaned clothing. As shown in Table II, the exposure levels are far lower

than for workers. The highest level exposures are for customers
using coin-operated laundry facilities in establishments that also
have coin-operated dry cleaning machines. Urban residents are
exposed to low ambient levels of PCE; the resulting exposure is
less than for users of commercial dry cleaning services.

Control Policies for Reducing PCE Exposure. The amount of perchloroethylene vapor escaping from machines can be reduced by a
variety of straightforward methods (18). Better maintenance of
machines, replacement of leaky gaskets and seals, and other
"housekeeping" measures might, on the average, reduce PCE losses
by 40% at little cost. In fact, these measures could result in an
annual saving of the order of $10 million for the industry from
reducing PCE purchases. A somewhat more costly option is the use
of a carbon adsorption unit to recover PCE vapor from the air in
the plant. Many plants already have these units, and if they were
used throughout the industry, PCE losses and worker exposure would
be reduced by an average of about 20%; exposure for service users
and urban residents would be reduced somewhat less. Because of a
credit for PCE recovery, the net cost of these units for commercial dry cleaners is very low. Even when coin-operated units are
included as well, the estimated net cost for using carbon adsorption units throughout the industry is only about $3 million
annually. Putting coin-operated cleaners in a separate room from
laundry facilities could reduce exposure for coin-op laundry
service users by about 90%, at an annual cost of about $5 million.
Finally, more expensive dry-to-dry machines could be used in
commercial cleaners, reducing machine operator exposure by a third
at a cost of $9 million annually.

Risk Assessment for Perchlorethylene. The above estimates for
exposure, exposure reduction under controls, and human cancer
incidence given exposure can be combined into a risk assessment.
A summary of the results is given in Table III. If the CAG
assumptions for the dose response relationship are used, the
projected cancer incidence is about 350 cases per year, the
majority of which occur in workers, with most of the remainder in
service users. The lifetime probability of cancer for a machine
operator is 23%, a high enough number that one would expect to see
strong epidemiological evidence if these assumptions were correct.
(Some epidemiological evidence does suggest an increased risk for
cancer among dry cleaning workers (19,20), but not an effect of
this magnitude.) If all of the control options discussed above
were implemented, expected cases of cancer would be reduced about
two thirds under the CAG assumptions, with a somewhat larger
reduction in incidence among service users than workers. We might
conclude, however, that even with these controls PCE would remain
a significant public health problem.
    If instead of the CAG assumptions we use the alternative
assumptions on the right of Table I, a very different picture

Table III.   Summary of PCE Risk Assessment

	CAG Assumptions (Present Exposure)	CAG Assumptions (Full Controls)	All Alternative Assumptions (Present Exposure)
Expected Number of Annual Cancers:	347	112	0.01
Workers	181	84	0.01
Service Users	163	26	10^{-5}
Urban Residents	3	1.5	10^{-8}
Lifetime Probability of Cancer:			
Machine Operator	0.23	0.08	3×10^{-5}
Coin-Op Laundry User	2×10^{-4}	1×10^{-5}	10^{-11}
Nearby Urban Resident	2×10^{-5}	1×10^{-5}	10^{-13}

emerges. The expected cancer incidence in that event is only one case per hundred years, a change of nearly five orders of magnitude. The lifetime probability of cancer estimate for a machine operator at 3×10^{-5} still is not negligible, because of the high level of the occupational exposure. The incidence and lifetime probability of cancer for service users and urban residents become negligible under the alternative assumptions.

Our report for EPA examines each of sixteen combinations of control options for each of the eight combinations of dose response assumptions. The resulting 128 scenarios correspond to the end points of the decision tree shown in Figure 1. For each of these 128 scenarios we worked out the impacts on workers, users, and urban residents in the same manner as shown in Table III.

Decision Analysis for Perchloroethylene.   Decision analysis provides formal methods for selecting the best control decision in the face of uncertainty on the dose response relationship. Two sets of inputs are needed: (1) judgmental probabilities describing the likelihood of the assumptions and, therefore, of the eight dose response cases considered and (2) a monetary equivalent value per case of cancer avoided so that health and economic impacts can be compared. While an important output of the analysis is the recommended control decision, this recommendation depends on the input judgments about the dose response uncertainty and the value of avoiding a case of cancer. Sensitivity analysis can demonstrate how changes in these judgments affect the recommendation on the control decision. The insights from sensitivity analysis are

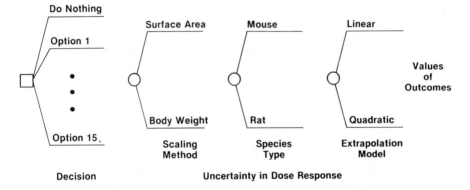

Figure 1. Decision tree for PCE control analysis. Key: □, decision; and ○, resolution of uncertainty.

usually the most important results from decision analysis. These
insights identify which judgments are critical in the selection of
the best decision alternative, and which judgments are less significant because over a wide range of values the same decision
alternative remains preferred.

We might illustrate by briefly summarizing the insights from
our illustrative calculations on perchloroethylene. As base case
assumptions, we used the value of one million dollars per case of
cancer avoided and probabilities of 20% to 50% for the CAG assumptions (implying 50% to 80% for the alternative assumptions listed
on Table I). For these probabilities the expected incidence of
cancer is a few cases to a few tens of cases per year, and the
most costly control alternatives are not judged worthwhile. (If
the three CAG assumptions were to be judged certain or very nearly
so, then the analysis would indicate that all of the controls
would be worthwhile, since the reduction in cancer incidence times
a million dollars then exceeds the annual cost for adding each of
the control options.) Among the control alternatives, the option
of better housekeeping and maintenance is clearly preferred to the
present situation because reduced PCE consumption provides net
economic gains and the health impacts to workers, users, and urban
residents are all reduced. Carbon adsorption units are the next
most attractive option, because their net cost is low and they
afford significant reductions in exposure and therefore potential
decreases in cancer incidence. Locating coin-op dry cleaning
machines in separate rooms may also be worthwhile if the probability of adverse dose response cases and the value of avoiding a
case of cancer are judged to be high enough.

Another important set of insights from decision analysis
comes from evaluating what it would be worth to resolve uncertainty before making a decision. By such means as larger scale
bioassays and pharmacokinetics research it might be possible to
resolve which of the eight sets of dose response assumptions is a
reasonable approximation to reality. How much should we be willing to pay to obtain such information? For the perchloroethylene
case study, our illustrative calculations show the value of the
information to be in the range of one to four million dollars per
year.

Insights and Conclusions

Is the risk of cancer posed by perchloroethylene "unreasonable" under the language of TSCA? No clear answer emerges from
the illustrative analysis. Whether a risk is unreasonable is not
a matter to be determined from scientific evidence on toxicity and
exposure, but rather a determination that will hinge on judgment.
We concluded from our calculations that the uncertainty in projected annual cancer incidence from PCE was nearly five orders of
magnitude, and such large uncertanties in health impacts may be
typical for many chemical agents.

The major insight from the perchloroethylene case study comes from the comparison between risks to the workers, the users of dry cleaning services, and the public that is exposed to low ambient levels of PCE in the air. If there is a significant incidence of cancer from PCE exposure, the effects will be predominantly among the workers rather than the users and the public. Calculations of the type we have carried out should be useful and illuminating not only to the regulatory agencies, but to individual company managements, workers, and consumers involved with a chemical agent. Many dry cleaning establishments are owned and operated by families, so that the management and the workers are the same people. Given that there is a suspicion based on animal bioassay evidence that PCE may induce cancer in humans, some dry cleaning plant owners may wish to reduce PCE exposure by means such as carbon adsorption units whether or not they are required by regulatory agencies to do so.

Our assignment for EPA was to apply quantitative risk analysis methods to the determination of risk for a particular chemical. The health risks for perchloroethylene turned out to be highly uncertain, but by using decision analysis concepts we were able to display this uncertainty in terms of alternative assumptions about the dose response relationship. Similar methods might be used to characterize uncertainties about human exposure to a chemical agent or about the costs to producers and consumers of a restriction on chemical use.

The methods of decision analysis provide a promising way to expand risk assessment and cost-benefit calculations to include uncertainty. The use of these methods in carrying out analysis for a specific chemical is not just a matter of crunching numbers through a formula; it requires skillful formulation of the analysis to reflect biological, economic, and other factors crucial to the regulatory decision. It may not be possible to establish a single number for the "probability of harm" as mentioned in the TSCA legislative history on the basis of hard scientific evidence; the "probability of harm" will usually be a reflection of scientific judgment, and the judgments of different scientists may often disagree. But probability is the right language for addressing the problem. An alternative to introducing probabilities is to make worst-case assumptions. Such assumptions may be useful in determining upper bounds on the health risks of chemical agents so that low risk chemicals can be eliminated as subjects for regulatory attention. But when the worst-case estimates are high, worst-case assumptions on the extent of health effect incidence may serve little useful purpose for regulatory decision making and only frighten people who find they have been exposed to the chemical agent in question. What is needed is a careful examination and synthesis of the scientific information available so that regulators, chemical companies, and the public can balance the probability of harm against the benefits to be lost if the chemical agent is controlled or restricted.

Risk assessment, cost-benefit analysis, and decision analysis do not provide an easy means of calculating the right answers for regulatory decisions under TSCA. These decisions are highly complex and uncertainties abound. What quantitative analysis can provide is a decision framework where the complexities and uncertainties can be set forth and examined by those with an interest in the decisions and the time and motivation to explore the issues in detail. To the extent that decision frameworks based on quantitative methods can provide insights and improve the process of communication and consensus building, they will have a useful impact in improving industry and government decision making under TSCA.

Literature Cited

1. "Legislative History of the Toxic Substances Control- Act," U.S. House of Representatives, Committee on Interstate and Foreign Commerce, 1976, p. 14.
2. Raiffa, H. "Decision Analysis: Introductory Lectures on Choices Under Uncertainty"; Addison-Wesley Publishing Co.: Reading, Massachusetts, 1968.
3. Brown, R. V.; Kahr, A. S.; and Peterson, C. R. "Decision Analysis for the Manager"; Holt, Rinehart, and Winston: New York, 1974.
4. Holloway, C. "Decision Making Under Uncertainty: Models and Choices"; Prentice-Hall: Englewood Cliffs, N.J., 1979.
5. Campbell, G. L.; Cohan, D.; and North, D. W. "The Application of Decision Analysis to Toxic Substances: Proposed Methodology and Case Studies"; prepared by Decision Focus Incorporated for the Office of Toxic Substances, Environmental Protection Agency, 1982.
6. National Research Council, "Decision Making for Regulating Chemicals in the Environment"; National Academy Press: Washintgon, D.C., 1977.
7. "Decision Making in the Environmental Protection Agency," National Academy of Sciences, 1977.
8. "Drinking Water and Health," National Academy of Sciences, 1977.
9. "Regulating Pesticides," National Academy of Sciences, 1980.
10. "Diesel Cars: Benefits, Risks and Public Policy," National Academy of Sciences, 1982.
11. "Risk and Decision Making: Perspectives and Research," National Academy of Sciences, 1982.
12. Albert, R. E.; et al., "The Carcinogen Assessment Group's Method for Determining the Unit Risk Estimate for Air Pollutants," U.S. Environmental Protection Agency, 1980.
13. Anderson, E. L. "Risk Assessment and Application to Carcinogen Policy: Development and Current Approaches," paper presented to the 184th National Meeting of the American Chemical Society, Kansas City, MO, 1982.

14.   Albert R. E.; et al., "The Carcinogen Assessment Group's Carcinogenic Assessment of Tetrachloroethylene (Perchloroethylene)," U.S. Environmental Protection Agency, 1980.
15.   U.S. Environmental Protection Agency, Science Advisory Board, Subcommittee on Airborne Carcinogens. Public meeting held in Washington, D.C. on September 4 and 5, 1980.   Transcript produced by Neal R. Gross, court reporters and transcribers, Washington, D.C.
16.   Guess, H; Crump, K; and Peto, R. "Uncertainty Estimates for Low-Dose-Rate Extrapolations of Animal Carcinogenicity Data," Cancer Research, 1977, 37, 3475-3483.
17.   Ludwig, H. R. "Occupational Exposure to Perchloroethylene in the Dry Cleaning Industry"; National Institute for Ocupational Safety and Health: Cincinatti, Ohio, 1981.
18.   McCoy, B. C. "Study to Support New Source Performance Standards for the Dry Cleaning Industry"; report prepared by TRW for the U.S. Environmental Protection Agency, 1976.
19.   Blair, A.; Drople, P.; and Grarrman, D. "Causes of Death Among Laundry and Dry Cleaning Workers," American Journal of Public Health, 1979, 69, no. 5, 508-511.
20.   Lin, R. S. and Kessler, I. I. "A Multifactorial Model for Pancreatic Cancer in Man," Journal of the American Medical Association, 1981, 245, no. 2, 147-152.

RECEIVED October 29, 1982

Meeting the Needs of TSCA

Educating the Environmental Chemical Professional

R. L. PERRINE

University of California, Department of Environmental Science and Engineering, Los Angeles, CA 90024

This article describes educational preparation particularly suited to the environmental chemical professional. As the decade of the 1970's brought growing concern for the environment, the UCLA program leading to the degree of Doctor of Environmental Science and Engineering was developed. Discussion of this program includes a rationale for education in the 1980's, discussion of the body of knowledge seen to underlie the professional field, and the curriculum by which knowledge is transferred to students. Unique elements include on-campus interdisciplinary project work, and a several year off-campus internship. Final discussion addresses accomplishments and difficulties experienced, and takes a look to the future.

The decade of the 1970's has seen mushrooming growth in concern for environmental health and the broader issues of the natural environment. This attitude is reflected in much legislation. Following incidents involving chemical pollutants, and recognition of chemical hazards and potential impacts of toxic substances, the Toxic Substances Control Act (TSCA) was added to prior law. Objectives of TSCA and its impacts on the chemical industry are addressed elsewhere within this symposium. The present paper addresses education to meet the mandates of TSCA.

TSCA poses a new challenge to universities. While toxics are uniquely a chemical problem, their impacts extend to involve other disciplines as well. The life cycle of a toxic substance starts with chemical feedstocks. It continues through the myriad steps of manufacture and processing, through use, and ends only after end-product disposal. During this cycle there are many chances for "leaks" into the environment. Risks posed may appear primarily as risks to occupational safety and health, to general human health, or to particularly sensitive or important

0097-6156/83/0213-0197$06.00/0

elements of some ecosystem. As a result it often will prove
necessary to understand processes extending throughout the bio-
sphere. Thus education to manage toxics must bring a number of
disciplines to bear in parallel on any problem; educational
needs are very broad, often interdisciplinary, but not beyond
our ability to perceive, to plan, and to execute effectively.

The remainder of this paper will address the broader preparation of the environmental professional which fits needs
derived from the several environmental legislative mandates, and
which by its nature is particularly well suited to the specific
needs of TSCA and the environmental chemical professional. The
basis for discussion is a graduate level program in the application of environmental science developed over more than a decade
at UCLA. Our experience suggests that if TSCA-like concerns are
to continue and to be met, educational approaches such as this
one will likely assume a growing importance.

A Characterization of Environmental Science Educational Needs

Representative toxic substance problems for which education
is likely to be needed range widely. Example challenges include
the fate of toxic and hazardous materials in the air environment, cost comparisons of treatment and disposal alternatives
for hazardous materials, persistence and movement through the
geohydrologic environment, effects of toxics on plants and animals, and so forth. Historically there has been no one profession both broad enough and focused to anticipate and understand
the interactions among the disciplines involved (1). Each field
has its quota of experts and their expertise can be called on to
evaluate some environmental phenomena. However, there has been
no expertise to draw together and integrate these fields.

By their nature, problems faced in environmental health,
the natural environment, and resources are societal problems;
the subject of public policy decisions. Experience suggests
that in such situations findings as to fact and consequences
must be established separately from and should precede policy
decisions insofar as possible, thus gaining objectivity. However, the scientist seeking useful input soon realizes that the
ultimate goal must be kept in mind at every step. It establishes the language in which results must be stated, deter-
mines priorities since problems greatly outnumber resources to
solve them, and sets depth of appropriate investigation.

Another essential is that the work of the environmental
professional is applied research; the use of more basic information generated by others. Different tools, philosophy and motivation are needed.

"Man seems to have at least two natural drives which
put the impetus behind science. One is the need to
understand; the second is, that once man understands
he wants to see what he can do with that understand-

ing. It's in the dichotomy that I see the difference between basic and applied research. Basic
research is done to <u>understand</u>; applied research is
done to <u>do</u>. The methods used may be the same. The
difficulty and sophistication of the work may be the
same. Both are important. Motivation is the only
difference." (2)
In response to this need the University of California, Los
Angeles, established a curriculum to educate the "Environmental
Doctor". It is (formally) an Interdepartmental Program in Environmental Science and Engineering, and leads to the degree of
Doctor of Environmental Science and Engineering. Its goals are
different than those of other doctoral-level, environment-related degree programs, thus it uses somewhat different tools and
educational environment. This presentation will document the
program basis and accomplishments, all with particular concern
for needs derived from TSCA.

Rationale for Education in the 1980's and Beyond

Needs in environmental education derive directly from the
dozens of problem areas which need to be addressed and which all
share certain charateristics. They are portions of a continuum. Study of a toxics problem may start at any point; for example, with the chemical process by which a toxic is first made,
or with impact on aquatic ecosystems. But societally useful
knowledge means learning about the entire continuum, not just
details of a specialized portion. Environmental education directed toward toxic substances must extend beyond a chemical
technology and its direct impact to the full range of impacts
and options -- what will make them "safe", their internal and
external costs, and the way they may fit into the fabric of
society -- all such knowledge must reach decision-makers. The
fact that such problems exist, and that past education has not
prepared a generation well to deal with them, leads to the rationale behind UCLA's Environmental Science and Engineering.
As noted by Wolman (3), educational institutions are con-
tinually asked to prepare those who will search for solutions of
societal problems. Problems in the real world do not separate
nicely into "disciplines". We do not see the "botany problem",
or the "meteorology problem", or the "chemical engineering problem", as such. Rather, we see a minor by-product from a facility designed by a chemical engineer. Released, it is trans-
ported by meteorological processes, and becomes of concern because a botanist foresees ecological damage as a consequence of
its downwind presence. Thus while disciplines and departments
in universities are an administrative convenience and provide a
perhaps needed foundation for specialized research and education, educational institutions also must address problems which
do not fit nicely into present disciplinary units.

For at least the last century science has been fission-
ing. Natural science has become physics, chemistry, mathematics
and biology. Biology has further split into botany, zoology,
molecular biology and microbiology. Even within microbiology
there is a separation into virology and bacteriology, and further distinction between disparate environments such as the
microbiology of soils and of fresh water aquatic ecosystems.
     Perhaps this pattern has helped good teaching and research
following the traditional reductionist approach (4). This approach -- in which a component of a field of study is isolated,
eliminating the influence of variables, and studied in depth --
no doubt has contributed to much of the incredibly rapid pace of
advance over the recent past (5). However, there comes a point
at which the organizational pattern and the accompanying learning experience no longer serves in an optimal fashion. A kind
of tribalism is reinforced by disciplinary jargon and by treat-
ing those who venture from the territory of the tribe (or who
intrude on it) as enemy aliens (5). Further, a pattern perhaps
useful for basic teaching and research gains a perhaps unearned
intellectual significance as departmental administrative units
certify themselves as "disciplines" (4).
     No high quality research investigation is likely to be en-
tirely free of the reductionist approach (5). On the other
hand, successful problem-oriented research cannot be carried out
in splendid isolation (4). There is a need to transcend reduc-
tionism as preparation for the major problems of our time, which
require that information be integrated so that a complex system
can be studied as a whole. An elegant analysis of this need has
been published by Odum (6). Students must be attracted to working on these problems and exposed to the holistic view essential
to their solution (5, 7). Without a deliberate effort along
these lines, problems will be attacked in bits and pieces as
graduating students -- clones of the faculty who have educated
them -- hack away at the miniscule, exposed portion of a problem
conveniently close to their graduate school specialty. Under
such conditions problem solutions are not likely.

The Origin of UCLA's Environmental Science and Engineering

     In 1969, with recognition of environmental problems emerg-
ing, the resources of the University were put to work for the
benefit of the state of California. Essential characteristics
of what was needed to resolve environmental problems became
apparent from work addressing California's critical concern: air
quality. These contrasted with basic research themes.
     At this point Dr. Willard F. Libby, Professor of Chemistry
and Nobel Laureate, stepped in. Dr. Libby had an idea -- perhaps not for the first time in his life -- but a truly heretical
idea among those grounded in basic science. What he envisioned
as essential to solve the inherently interdisciplinary problems

of the environment was a <u>clinical kind of preparation</u>. Well-grounded people, steeped in the basics, did not in his view need ever longer and more close range inspection of that discipline-limited dot in the universe that represents the usual Ph.D. dissertation. No matter how skillfully executed, most such effort was not sufficiently broad-ranging to strike any real target, and so was ineffective in solving problems.

What Dr. Libby at this time properly perceived as missing was exposure to the <u>real world of problem-solving</u>; hence the need for a clinical kind of preparation. Thus the concept he proposed was to create the "Environmental Doctor" -- the competent generalist, reasonably skilled in all aspects of the environment and expert in some, with the perception and judgement to select those few critical parts of a problem essential to its solution, and the management skills to assemble a team to perform the actual solution effort.

This, then, became the task of a small core of UCLA faculty: to create an actual academic program which, by contrast with chance or the slowly accumulating scars of experience, would efficiently prepare people for the transition from the idealized, contemplative world of the university to the harsh and often political realities of environmental problem-solving. Program ideas were tested and the concept took on a recognizable structure. A program existed, even if in embryo form and only as a "bootstrap" operation, largely fueled by the after-hours effort of a number of the principals.

The Body of Knowledge and the Curriculum

The structure of the university, and its traditional delegations of authority and responsibility, are designed to assure the orderly transfer of knowledge to students who are then awarded a degree. Environmental Science and Engineering prepares professionals for environmental problem-solving by participation in a clinical interdisciplinary curriculum. Thus it differs substantially from related but conventional Ph.D. Programs. Our objective is to develop a high level of skill at the application of knowledge. To do so requires a delicate balancing of emphasis: a sufficient basic depth, plus a useful level of competence across a mix of disciplines.

Breadth of interest makes it quite likely that interdisciplinary program activities will tread on the toes of others. Territorial jealousies, and the fact that there is not yet well-defined, demonstrated theory or methodology for much of the work to be done combine to ensure a degree of controversy (<u>8</u>). Thus it is essential to ask what is the nature of learning appropriate to the degree, and how a proper level of achievement can be assured, and to find a satisfactory answer.

<u>Body of Knowledge</u>. (These sections derive from working

documents developed by the UCLA Environmental Science and Engineering Interdepartmental Committee as a response to recommendations of the Graduate Council growing from a six-year review of the Program.) The body of knowledge which should underlie the program is defined by the nature of the profession. Graduates can expect to be employed to assess impacts of alternative courses of action on the environment and resources, to recommend sound policy, and to devise means to implement policy once a decision has been reached. Typically, such activity will require quantitative synthesis of information from several traditional academic disciplinary fields. Towards this end, students must achieve a broad understanding of the environment and resources and acquire technical and integrative skills enabling them to function at the highest levels of responsibility.

Graduates are employed in technical assessment and management positions with governmental agencies, consulting, and industrial firms concerned with environment-related projects. Their rapid rise to relatively high-level positions is felt to be a result of a societal need for scientists with the advanced interdisciplinary training provided. The present focus, interdisciplinary training in the environmental sciences and their application, is a successful one. We see no reason for major change. This training has been met in the curriculum through courses, case studies, and problem solving opportunities.

Based upon evaluation of the more useful courses and the views of faculty, students, and graduates, the necessary body of knowledge associated with the degree can be defined and organized under three broad topics: environment, environment and technology, and environment and society. Subject matter is presently taught in courses in departments spread over the university. Often the most useful content is presented at an undergraduate or beginning graduate level, in serious but general courses designed for departmental majors.

The knowledge required under environment should be a thorough understanding of the characteristics of terrestrial, air, and water environments; of the biota; and of geological, biological, chemical, hydrological, and meteorological processes. This is a considerable body of knowledge. However, less is actually required of our graduates: specific knowledge of those characteristics and processes of the environment that are subject to disturbances (such as pollution), that represent resources that can be exploited, or that can serve as impediments to man's activities, together with the fundamental principles required to understand such processes.

Knowledge required under environment and technology is a compendium of the technical and analytical tools necessary to solving environmental problems, or to developing technology and policy that can avoid them. Required emphasis is on energy technology, particularly new energy sources; pollution control technology; environmental measurement, modeling, and analysis;

the characteristics and sources of pollutants; and the pathways through which pollutants impact human health and the environment. The level of knowledge should be such that the individual could, in principle, use all necessary analytical tools. However, the major emphasis in the curriculum should be on developing a knowledge of the appropriateness of various methods and technologies, their strong points, and their shortcomings.

The knowledge associated with environment and society relates to the social and institutional factors relevant to environmental problem-solving. Emphasis should be on methods for assessment of social and economic impacts, legal constraints and processes, and implementation of policy.

Clearly, the total body of knowledge as presented above exceeds what might realistically be mastered in the first two years of preparation for the Doctor of Environmental Science and Engineering. However, considerable selectivity is possible based upon the needs and previous background of an individual. As an example, a thorough understanding of environmental toxicology, the chemistry of toxics and methods for their destruction, and related regulations would reasonably offset a superficial knowledge of air pollution.

From our experience, further refinement to establish a suitable minimum under the three topics above appears best accomplished by doctoral faculty committees selected for the individual, considering individual needs. The basic body of knowledge might be conceived as an understanding of the environment, its characteristics and processes, a knowledge of environment-related technology, an ability to use quantitative analytical techniques, and an appreciation for the social and institutional framework of environmental policy.

Requirements at Entrance and the Curriculum Framework. The curriculum is rather simple and straightforward. Formal entry requires a Master's degree in a field within the natural sciences, engineering, or public health. Preferably the Master's would include a strong, independent thesis effort. The intent of this requirement is to insure that the student have (and retain) competence within an established discipline. Students are carefully selected from applicants on the basis of prior performance, test scores, recommendations, and interviews, and thus with a goal of selecting synergistic combinations of intellect, aptitude, and motivation likely to lead to success.

In the four-year UCLA program, students successfully:
o   take additional courses in areas peripheral to the student's specialty in order to obtain the breadth necessary to successfully work on problems inherently interdisciplinary in nature,
o   take additional courses in the Master's area as judged necessary to establish and retain an appropriate level of disciplinary competence,

o   take and pass a series of cumulative exams that stress
     current awareness and ability to respond,
o   spend a year at the university as a member of an interdis-
     ciplinary team participating in an intensive problem-sol-
     ving experience,
o   take an oral examination for advancement to candidacy,
o   spend several years as an intern at an outside institu-
     tion, gaining applied research experience under guidance,
o   demonstrate acquired competence during a one-term return
     period at UCLA, and
o   prepare written and oral reports to document the applied
     research experience for deposit in the archives at UCLA.

     Course Preparation. No specific courses are required, but
there are suggested courses through which requirements can be
met. Subject matter is conveniently organized under the three
broad topics as developed in the prior section.

                  TABLE I. BREADTH COURSES

                      The Environment
Environmental Chemistry   Environmental Geology
Air Pollution   Water Pollution
Hydrology   Oceanography
Meteorology   Ecology
Soil Science   Microbiology
                  Environment and Technology
Air Pollution Control   Water Pollution Control
Energy Resources and Technology Risk Assessment
Microbiological Control   Environmental Health
Environmental Toxicology   Occupational Health and Safety
Environmental and Pollution   Environmental Measurement
  Modeling
                  Environment and Society
Environmental Law   Environmental Impact Assessment
Environmental Policy   Environmental Regulation
  Implementation   Environmental Planning and
Resource Economics   Management

     Not all students take exactly the same program of cour-
ses. An individual's curriculum will be determined with the
approval of his or her guidance committee and the graduate advi-
sor. Areas of concentration, usually growing from an anticipated future, are encouraged.
     For many subjects present courses in traditional depart-
ments deal adequately with needed content. For other subjects
specially developed content is essential. Presently, students
have been meeting the requirement for knowledge by about 5 selected courses that could be listed under the heading of environment, 4 to 6 under environment and technology, and 4 under

environment and society. It is anticipated that there will be a
modest growth in courses developed specifically for Environmental Science and Engineering, primarily to ensure that essential
content is not neglected. Prerequisites include mathematics
(calculus through differential equations and statistics), chemistry (through organic), biology, and earth science courses.
All are preferably part of preparation prior to admission.

Literature as well as lecture courses can serve as a basis
for the required body of knowledge and hence student preparation. A vigorous effort is made to encourage regular and current reading of the literature. The subject area is vast, and
a periodicals bibliography is very useful.

Problems Courses. The interdisciplinary team problem-solving experience -- the UCLA "Problems Courses" -- are a unique
and essential part of the curriculum. Usually about three faculty from different backgrounds provide guidance, along with
post-internship stage students, to perhaps six second-year doctoral students in a saturation teaching environment. The rationale is that leaders of problem-solving teams of the future
should experience as early as possible the rigors of addressing
open-ended problem statements and real-time decision-making.
They also learn the demands that up-to-date, innovative use of
more basic research places on problem solvers. Thus these unique courses require students to quantify and measure necessary
parameters, perform critical evaluation, and edit and process
technical and socioeconomic information. Finally, they require
the effective communication of study results through a final
report on a complex, policy-related subject, both to the competent lay-person and the technical specialist.

Subjects have included almost the full range of possible
topics. In each case study, results have been pepared as a
formal Environmental Science and Engineering report. A list of
reports and sponsoring organizations is included as an appendix.

The benefit of problems courses is not limited to enrolled
Environmental Science and Engineering students. As noted, environmental problems always involve policy, and this demands deep
involvement of the social sciences. We have achieved this
largely through participation drawn from relevant non-science
disciplines, and thus a truly interdisciplinary team effort.
The deeper probing made possible by an effort shared with basic
science and engineering also is encouraged. A particular value
to such efforts derives from the fact that realities of an un-
even data or knowledge base frequently limit the problem-solving
that can be accomplished. Transferring knowledge of the problem
may encourage needed research in depth.

Internships. Much like problems courses, internships represent a unique element of the curriculum. Internships have
proved easy to arrange and to monitor; participating institu-

tions are included as an appendix. Interns each arrange for
their own position though with a great deal of assistance from
faculty, and (with a maturing program) from graduates and other
interns. Internships are paid positions, with each individual
expected to earn his keep. The qualities we seek in an internship are threefold:

o   a position which will challenge and provide responsibility
     for the intern,
o   suitable interdisciplinary character to the problems ad-
     dresred,
o   an organization with an earned reputation for leadership,
     and which regularly addresses the more difficult and chal-
     lenging of environmental problems.

Interns are regularly visited by faculty who discuss their work
and progress with them and their supervisors, and maintain a
continuing oversight until the internship requirement is met.
Interns also submit quarterly reports on their progress.

    Many internship positions have focused directly on toxic
substances. Generally these have built on a particularly well
suited undergraduate and Master's-level background, plus the on-
campus broadening experience of the doctoral program, to select
an opportunity for growth through toxics-related internship work
which should lead to a productive, life-long career. Specific
internship projects have addressed biological monitoring for
toxics in aquatic environments, the establishment of chemical
test methods particularly suited to environmental monitoring
needs, occupational health risks, transport and bioaccumulation
of toxics in the terrestrial environment, disposal methods for
specific compounds and for classes of toxics, and regulatory
needs to meet the requirements of legislation.

    Perhaps the greatest value of the internship is that stu-
dents often gain the opportunity to work on truly significant
problems. On campus, their efforts would be limited by the
ability of the campus infrastructure to address problems in a
real rather than artificial manner. Arrangements would be much
less flexible, and less readily adapted to meet the emerging
needs of the individual. True, an on-campus base might always
be seen to involve less risk and offer greater administrative
convenience. However, in our experience, there is no preferred
substitute for actual experience.

Concerns of Particular Importance

    An innovative program, even while seeking solutions to
recognized problems and achieving substantial success, raises a
host of concerns in the minds of faculty and administrators.
These must be openly addressed. Experience with other interdisciplinary programs has evoked similar concerns, and similar
experience with their resolution (3, 4, 5). Brief overviews of
primary areas of concern are presented here.

Program Philosophy: To What Degree Should the University
Respond to Society's Needs?   There is a very basic question as
to whether or not the university should directly prepare individuals to solve society's problems.   A point of view is that the
university does basic research and teaching well, with few alternatives available.   It is seen as unwise to attempt to extend
limited resources to address applied topics, where private and
governmental agencies are active.   It is argued that the university should provide a quality education in depth.   If breadth
should then be needed in a subsequent career, the argument continues that further preparation and experience can be obtained
on the job.   Wolman, though not in agreement with this thesis,
has summarized the philosophical viewpoint succinctly: "the
best generalist is a broken-down specialist" (3).
    It is clear that those who support the UCLA Environmental
Science and Engineering approach and many others do not agree.
One reason is that attitudes developed early tend to persist
throughout one's career (5).   Thus those who as students have
been deprived of a holistic view may strive ineffectually toward
inappropriate and narrow goals, wasting scarce resources, and
too-often remaining unaware of the greater accomplishment that
might have been.   This viewpoint is supported by evaluations
such as that (regarding chemical engineers) of Metzner (9): "We
are producing too few employable Ph.D's.   Further, their education is frequently too narrow to quality them for the salaries
which would make this degree economically attractive..."

    Curricular Questions.   There is a serious question as to
what represents the proper balance between depth and breadth.
An infinite variation in ratios is possible.   Tension between
advocates of different positions may always be a fact of life
for any interdisciplinary program.   If UCLA experience is a
worthwhile guide, different proportions, but all in a middle
band and built on a solid Master's-level foundation, may prove
appropriate for individual students and their ultimate careers.

    Organizational Structure and Support Base.   Interdiscipli-
nary efforts require cooperation beyond the usual unit bounda-
ries.   Mutual interests of diverse faculty must be brought into
convergence.   Perhaps more important, some administrative "home"
must be found.   No doubt a variety of alternatives, including
unconventional ones, could prove suitable.   For the past 20
years supradepartmental organizations have held a very important
position in research.   However, the traditional unit for teaching in the American university remains the department.
    As noted by Roy (4), the world of the university generally
presents a rather unimaginative picture.   Reasons should be
obvious.   "Disciplines" are part of a continuum.   In fact, separation is sometimes difficult.   Portions of chemical and civil
engineering, both addressing health-protective control techno-

logy, provide an example. Since no truly fundamental definition
can be made, an operational one is used. With which units is
"discipline X" connected? The answer is the "department of
X". Thus a generally accepted "discipline" results when enough
institutions establish roughly similar departmental administrative units. There is then a circle of recognized peers, able
(and most willing) to compliment each other's scholarly discove-
ries. It is this circle that is essential for favorable review
and thus the promotion of capable individual faculty.

Many arrangements are possible for an innovative, interdisciplinary program. A likely yet unfortunate response to changes
within the campus establishment may be similar to that from the
surgical insertion of a foreign organ into a human body –– a
serious effort by the host to reject the intruding entity, regardless of its potential value or even necessity. At this
juncture there is an absolutely critical need for academic
statesmanship. The principal burden will fall on academic administrators: they will need to demonstrate leadership, to
devise creative organizational relationships, and they must
provide the essential resources (despite shortages everywhere).

Faculty Relationships. Faculty selected to guide a program
and provisions for their future are critical to success. Qualified potential faculty are likely to ask several very important
questions. How will promotion and tenure be decided? Will
leaving established disciplinary departments aligned with professional and scholarly societies jeopardize their future? For
continuity, a core of faculty must be acquired and held together. These must not only cover the necessary range of expertise, but also have a continuing stake in the program's success
–– allied but independent and non-responsible departments are
all too likely at some future time to find that their current
best interests lie elsewhere, and withdraw "firm" commitments.
Thus it must be possible to successfully bring in qualified
faculty, and give them the autonomy, accountability, and rewards
needed for them to ensure the health and vigor of the program.

Research done is likely to be judged by traditionalists as
not falling within the mainstream of a discipline, and not suited to its prestige journals. Therefore it is said to be of
diminished quality. In point of fact, traditional discipline-
oriented faculty have no basis for judging interdisciplinary
research and often make no substantive effort to become informed
(5). Thus it becomes particularly important for interdisciplinary groups to have their own departmental stature with regard
to all matters of tenure and promotion.

Student Relationships. Establishing and retaining the
identity of students in an interdisciplinary program can be a
concern, deriving from reward structure, recognition, jobs,
etc., in the outside world. If our experience is a useful

guide, the student entering an interdisciplinary program of recognized high quality at a recognized institution is unlikely to face serious problems of identity in the future. On occasion it may be necessary to explain how and why the individual's particular preparation to become an "Environmental Doctor" came about and just what the qualifications actually mean. The greatest nuisance comes from some personnel officers and the simplicity of computerized thinking. They totally discourage a response not tailored to convention and the past.

An initial strong argument in support of requiring a Master's at entrance was retention of a recognized disciplinary foundation. In fact, this was thought of in part as security against future failure in the interdisciplinary world. No such "backstop" utilization has been necessary. Experience shows, however, that each individual needs to have experienced independent, intellectually demanding work at least at this level. There are also ties to professional societies and journals which should retain value throughout one's career.

UCLA students in Environmental Science and Engineering have established on-campus student organizations and an organization of program graduates. A number are active in emerging professional organizations such as the National Association of Environmental Professionals. The most valued relationship, however, is our well-established "family": program graduates and interns, other UCLA graduates who have worked on applied research projects with us, and including the employers of our graduates.

The Anticipated Future

By any pragmatic measure, the "Environmental Doctor" concept has become an established success. The future should appear bright. Unfortunately, concerns regarding acceptance of applied research and innovative education noted earlier apply in fact at UCLA as well as generally. There is a basic policy question which each institution must answer, and for which to obtain a forthright answer is not easy. The question is: where do we place the balance point in a choice between education that most conveniently adapts to the needs for understanding and research publications of university faculty, and education designed to meet the needs of society? One would hope that in a future which, despite temporary remissions, is certain to be constrained by resources and ecology, the balance point will permit survival of a healthy share of each. If so, the kind of program represented by Environmental Science and Engineering at UCLA will grow.

Literature Cited

1.   Libby, W. F. "The Profession of the Environmental Doctor: Five Years Old." Bioscience 1976, 26(12),751-752.

2.   Hubbard, H. M. quoted in Chemtech 1979, 9(9),566.
3.   Wolman, M. Gordon. "Interdisciplinary Education: A Con-
     tinuing Experiment." Science 1977, 198(4319),800-804.
4.   Roy, Rustum. "Interdisciplinary Science on Campus - the
     Elusive Dream." Chem. and Engr. News 1977, 55(35),28-40.
5.   Cairns, John Jr. "Academic Blocks to Assessing Environ-
     mental Impacts of Water Supply Alternatives." Proc. of
     the Thames/Potomac Seminars 1979, Anne M. Blackburn, Ed.,
     Interstate Commission on the Potomac River Basin; p 77.
6.   Odum, Eugene P. "The Emergence of Ecology as a New Inte-
     grative Discipline." Science 1977,195(4284),1289-1293.
7.   McCormick, J. Frank; Barrett, Gary W. "Ecological Manpower
     and Employment Opportunities." BioScience 1979,29(7),419-
     423.
8.   White, Irvin L. "Interdisciplinarity." The Environmental
     Professional 1979, 1,51-55.
9.   Metzner, A.B. "Projected Needs for Chemical Engineers in
     the 80's." Chem. Engr. Prog. 1980,76(10), 20-27.

Appendix I.   Environmental Science and Engineering Applied Research Reports

AIR POLLUTION AND CITY PLANNING -- FINDINGS, RECOMMENDATION,
   EXPLANATION, also RESEARCH INVESTIGATION,   Case Study of a
   Los Angeles District Plan. (1972) (Sponsor: U.S. EPA)
FACING THE FUTURE;   FIVE ALTERNATIVES FOR MAMMOTH LAKES, Report
   and Summary. (1972) (Sponsor: National Science Foundation)
ENVIRONMENTAL, TECHNICAL, LEGAL AND SAFETY ASPECTS RELATED TO
   FLOATING NUCLEAR POWER PLANTS OFF THE COAST OF CALIFORNIA.
   (1973) (Sponsor: National Science Foundation)
WATER QUALITY AND RECREATION IN THE MAMMOTH LAKES SIERRA, Report
   and Summary. (1973) (Sponsor: National Science Foundation)
FUTURE ALTERNATIVES FOR THE SANTA MONICA PIER. (1973)
   (Sponsor: National Science Foundation)
SURFICIAL AND ENGINEERING GEOLOGY OF PART OF THE MAMMOTH CREEK
   AREA, MONO COUNTY, CA. (1973) (Sponsor: National Science
   Foundation)
MODELING LOS ANGELES PHOTOCHEMICAL AIR POLLUTION. (1975)
   (Sponsor: Dreyfus Foundation)
POPULATION AND ENERGY IN LOS ANGELES; THE IMPACT OF DIFFERENT
   RATES OF GROWTH ON TRANSPORTATION, AIR QUALITY, HOUSING AND
   OPEN SPACE. Substudies include: TRANSPORTATION; IS THERE A
   CHOICE? IMPACT OF THE ENERGY CRISIS AND ESTIMATES OF FUTURE
   AIR QUALITY.   THE CHANGING PATTERNS OF HOUSING DISTRIBU-
   TION.   RECREATION DEMAND IN THE SANTA MONICA MOUNTAINS IN
   1990. (1975) (Sponsor: National Science Foundation)
WILDERNESS WATER QUALITY; BISHOP CREEK BASELINE STUDY, 1974.
   (1975) (Sponsor: Water Resources Center)
WASTE NUTRIENT RECYCLING USING HYDROPONIC AND AQUACULTURAL
   METHODS. (1975) (Sponsor: Rockefeller Foundation)

NON-POINT SOURCE WATER QUALITY MONITORING, INYO NATIONAL
FOREST, 1975. (1976) (Sponsor: U.S. Forest Service)
THE SULFATES PROBLEM: ITS EFFECTS ON THE ENVIRONMENT AND MAN.
(1976) (Sponsor: Dreyfus Foundation)
SOUTHERN CALIFORNIA OUTER CONTINENTAL SHELF OIL DEVELOPMENT:
ANALYSIS OF KEY ISSUES. (1976) (Sponsor: Ford Foundation)
UTAH COAL FOR SOUTHERN CALIFORNIA ENERGY CONSUMPTION. (1976)
(Sponsor: Scaife Family Charitable Trust)
STUDY OF ALTERNATIVE LOCATIONS OF COAL-FIRED ELECTRIC GENERATION
PLANTS TO SUPPLY ENERGY FROM WESTERN COAL TO THE DEPARTMENT
OF WATER RESOURCES. (1977) (Sponsor: California Department
of Water Resources)
BACTERIAL WATER QUALITY IN WILDERNESS AREAS. (1977) (Sponsor:
Water Resources Center)
AN ASSESSMENT OF ELECTRIC POWER GENERATING OPTIONS FOR THE STATE
OF CALIFORNIA. VOLUMES I AND II. Report and Summary.
(1978) (Sponsor: California Energy Commission)
POWER PLANT SITING ASSESSMENT METHODOLOGY. (1978) (Sponsor:
Electric Power Research Institute)
DISPOSAL OR STORAGE OF COAL GASIFICATION WASTES IN SOUTHERN
CALIFORNIA. (1979) (Sponsor: Scaife Family Charitable Trust)
INSTITUTIONAL BARRIERS TO WASTE WATER REUSE IN SOUTHERN
CALIFORNIA. (1979) (Sponsor: Office of Water Research and
Technology)
WORKER HEALTH AND SAFETY IN SOLAR THERMAL POWER SYSTEMS.
Substudies include: OVERVIEW OF SAFETY ASSESSMENTS. DATA
BASE AND METHODOLOGY FOR THE ESTIMATION OF WORKER INJURY
RATES. THERMAL ENERGY STORAGE SYSTEMS. ROUTINE FAILURE
HAZARDS. OFF-NORMAL EVENTS. SOLAR PONDS. (1979) (Spon-
sor: U.S. Department of Energy)
CALIFORNIA'S NORTH COAST WILD AND SCENIC RIVERS: ANALYSIS OF
INTER-AGENCY PLANNING AND TECHNICAL ISSUES (1980) (Spon-
sor: Ford Foundation)
ECOLOGICAL AND INSTITUTIONAL FACTORS IN COASTAL SITING OF A
COAL-FUELED POWER PLANT AT ORMOND BEACH, CALIFORNIA.
(1980) (Sponsor: Southern California Edison Company)
ENVIRONMENTAL PLANNING FOR NEW TOWNS; EXPERIENCE AND SELECTED
OPPORTUNITIES. (1980) (Sponsor: Royal Commission for
Jubail and Yanbu, Saudi Arabia)
SITING OF AN INTERNATIONAL FACILITY FOR STORAGE OF VITRIFIED
RADIOWASTE. (1981) (Sponsor: Electric Power Research In-
stitute)
COMMUNITY APPLICATIONS OF SMALL SCALE SOLAR THERMAL ENERGY
SYSTEMS. (1981) (Sponsor: U.S. Department of Energy)
ENVIRONMENTAL CONSIDERATIONS IN SITING A SOLAR-COAL HYBRID POWER
PLANT. Substudies include: ENVIRONMENTAL ASSESSMENT. AIR
QUALITY AND METEOROLOGICAL IMPACTS. (1981) (Sponsor: U.S.
Department of Energy)
THE POTENTIAL PRODUCTION OF AIR POLLUTANTS NEAR STPS RECEIVER
SURFACES. (1981) (Sponsor: U.S. Department of Energy)

OIL AND GREASE IN STORMWATER RUNOFF. (1982) (Sponsor:
Association of Bay Area Governments, Oakland, CA).
EVALUATION OF GREAT DESERTS OF THE WORLD FOR PERPETUAL
INTERNATIONAL RADIOWASTE STORAGE. (1982) (Sponsor: Electric
Power Research Institute)

Appendix II. List of Internship Organizations

The Aerospace Corporation
ANCO Engineers
Association of Bay Area
    Governments
Battelle Memorial Laboratories
Bechtel Corporation
Boeing Engineering and
    Construction
Booz-Allen Applied Research
(U.S.) Bureau of Land
    Management
California Air Resources Board
California Department of
    Health Services
California Energy Commission
California Water Resources
    Control Board
Committee on Resources, Land
    Use and Energy, State of
    California Assembly
Congressional Research
    Division, Library of
    Congress
Dames and Moore
EBASCO Services
Electric Power Research
    Institute
Environmental Resources Group,
    Jacobs Engineering
Environmental Science
    Associates, Inc.
Eureka Laboratories
Florida Solar Energy Center
Form and Substance
IWG Corporation
James M. Montgomery,
    Consulting Engineers
Jet Propulsion Laboratory
KVB Engineering
LA/OMA Project, Los Angeles
    County

Lawrence Livermore Laboratory
Los Angeles Department of
    Water and Power
(U.S.) National Bureau of
    Standards
National Institute of Occupa-
    tional Safety and Health
Northern Energy Resources
    Company
Oak Ridge National Laboratory
Office of Planning and
    Research, State of
    California
Office of Technology
    Assessment, U.S. Congress
Pacific Environmental Services
Project Concern International
    (The Gambia)
The Ralph M. Parsons Company
Regional Water Quality Control
    Board, Central Valley
    Region, California
Republic Geothermal
Research and Development
    Associates
Rockwell International
Science Applications, Inc.
Scandpower (Norway)
Socioeconomic Systems
Southern California Edison
    Company
Systems Applications, Inc.
Technology Service Company
TRW, Inc.
U.S. Environmental Protection
    Agency
U.S. Geological Survey
U.S. Navy (Energy and
    Environmental Technology)
Wright McLaughlin Water
    Engineers

RECEIVED August 23, 1982

Overall Costs and Benefits

J. CLARENCE DAVIES

The Conservation Foundation, Washington, DC 20036

The paper describes what progress has been made under TSCA in obtaining information about chemicals and in controlling chemicals that may pose an unreasonable risk. It also describes the direct and indirect costs of the act and explains why a quantitative comparison of costs and benefits is not possible. It suggests several ways to increase the benefits of the act and lower its costs.

In this brief essay I would like to do several things. I shall begin by explaining why the costs and benefits of TSCA cannot be calculated in quantitative, much less monetary, terms. I will then try to review what I perceive to have been the costs and benefits of the major sections of the act. Finally, I will propose a set of more general observations about the act which may point the way towards increasing the benefits and reducing the costs of toxic substances control.

Cost-Benefit Analysis

It is impossible to express the benefits and the costs of TSCA in dollar terms and then compare them. In saner times it would not be necessary to belabor this point, but there are now people who apparently will not get out of bed in the morning unless they are convinced that the dollar benefits of eating breakfast outweigh the dollar costs of removing the blankets.

Cost-benefit analysis is a useful analytical tool that in many cases can help decision makers to know what factors are involved in a decision and in some cases can give the decision maker some idea of the relative importance of the factors. It cannot do more than that in the realm of policy making.

There are a variety of reasons why cost-benefit analysis is so limited. It assumes certain judgments, such as the judgment that the existing distribution of purchasing power is

0097-6156/83/0213-0213$06.00/0

ideal, which some of us are not comfortable assuming. It ignores certain essential aspects of decisions, such as who will bear the costs and who will gain the benefits, which cannot and should not be ignored. Its results are in most cases determined by assumptions which are inherently somewhat arbitrary, such as what discount rate is used. And finally, for the kinds of public health decisions discussed in this paper, there is never enough information to provide the kinds of numbers needed for cost-benefit analysis.

The lack of information is well illustrated by trying to examine the costs and benefits of TSCA. The major benefits of the act are the adverse health effects avoided by whatever actions are taken under the act's authority or because of the act's existence. But for many actions, such as voluntary testing by industry, we are not sure whether to attribute the action to TSCA's existence. We also are not sure how to relate such actions to reduced exposure to potentially hazardous chemicals. Insofar as exposure is reduced, we usually do not have any precise idea of the health consequences of such reduced exposure. And even if we knew the health consequences we would not know how to place a dollar value on them.

The range of uncertainty can be partially illustrated by a case study that The Conservation Foundation did last year of the use of Tris on children's sleepwear. Our purpose was to get some fix on quantifying the benefits of the TSCA PMN program. The number of cancer cases attributable to Tris was conservatively estimated to range between 12,600 and 25,200. Assuming 12,600 cases, $100,000 as the value of a human life, and a discount rate of 10% compounded annually, the dollar value of the Tris-caused cancers was $170 million. However, if we assume 25,200 cases, $ 1 million as the value of a human life, and no discount rate, the dollar value is $25 billion, a difference of more than two orders of magnitude. An even wider range of assumptions, producing an even greater disparity of final results, could plausibly have been used.

Information on TSCA costs is not much better than on benefits. It has been argued that the greatest cost of TSCA is the inhibiting effect it has on the development and manufacture of new chemicals. The information we have on this effect consists of several estimates of the decline in the number of new chemicals marketed since TSCA's passage. In my opinion these estimates are quite unreliable. But even assuming that they are reliable, we do not know anything about the dollar value or the economic importance of the chemicals not marketed. We assume that, whether measured by dollar sales or production volume or originality of the chemical, the chemicals not marketed were less important than the chemicals marketed. But we do not know whether the difference is large or small. Technically, to measure the dollar cost to society of TSCA's adverse impact on innovation, one would have to add the hypothetical producer's surplus of the chemicals not marketed (a number which cannot be calculated) to the hypothetical consumer's surplus of the chemicals not marketed (a number which cannot be calculated) and subtract from this sum the total of the producer's and consumer's surplus of the existing chemicals that would have been replaced by the new chemicals (a number which also cannot be calculated). In short, the cost side of TSCA is no more calculable in dollars than is the benefit side.

I have gone through this discussion to dispel any expectations that I might tell you what the quantitative ratio of TSCA costs to TSCA benefits is. I hope I have also suggested that any regulatory analysis that gives such ratios for any aspect of toxics policy belongs in the category of what Mr. Reagan calls "government waste and fraud." But of course the lack of information is a problem that plagues any kind of attempt to evaluate TSCA.

Benefits of TSCA

The two major goals of TSCA, and thus the two major types of benefits it has for society, are preventing unreasonable risk from chemicals and obtaining adequate information about the risks of chemicals. These two goals are inextricably related. There is little benefit in obtaining adequate information unless one is prepared to act on the information and, conversely, action to prevent unreasonable risk depends on the availability of adequate information. Since by definition it is not worthwhile to try to separate things that are inextricably related, I will discuss the benefits of TSCA primarily in relation to new and existing chemicals, rather than in relation to information and risk prevention.

New Chemicals. There are several ways one might try to estimate the benefits of the TSCA pre-manufacturing notification (PMN) program. None of them are satisfactory.

First, one could observe that in the 2-1/2 years that the PMN program has been in effect there have been no known cases of people dropping dead because of a new chemical. This is not a very informative observation, however, because it is more a comment on the long latency period of chronic effects and our inability to detect chemical problems than it is an observation about the lack of problems with new chemicals. This is not to say that chemicals with adverse effects have gone through the PMN process--only that we would not know it if they had.

A second approach to measuring benefits would be to look at changes that have occurred because of formal action or the threat of such action under TSCA Section 5. Of the more than 1,000 PMNs submitted since July 1979, nine have been subject to Section 5(e) orders and several others were withdrawn before a

5(e) order was prepared. According to the Office of Toxic Substances, EPA "has successfully negotiated voluntary controls or further data" for approximately 60 additional chemicals. (1, p. III-3) We do not know whether the dozen or so chemicals withdrawn from the PMN review would have posed an unreasonable risk or, even if they had, whether the risk would have been any more worrisome than the chemicals they would have replaced. The nature of the voluntary measures taken on the 60 chemicals is lost in the murky waters of confidential information, so there is no way to ascertain the benefits of these actions either.

A third benefits measure would be the extent and nature of voluntary actions taken by manufacturers. We have no information or even basis for guessing whether new chemicals that would have been manufactured five years ago are now voluntarily shelved because of potential risks to health or the environment. Certainly there is greater awareness of such considerations within most firms and also a greater ability to do toxicologicl and other relevant testing within many firms. It is not clear, however, what the testing resources are used for. The absence of data submitted with the PMNs indicates that the testing is not done on new chemicals before the PMN stage. Whether testing is done at a later stage in the development of a new chemical, is done mostly on existing chemicals, or is not done at all on the company's own products is not clear to me. Carl Umland of Exxon is chairing a committee for the Chemical Manufacturers Association to develop measures of voluntary compliance with TSCA. We badly need such measures, but it is important that they be credible and informative and not lend themselves to charges of being industry propaganda.

To summarize the benefits of TSCA with respect to new chemicals, there is a bit of evidence that indicates that unreasonable risk may have been averted from a few new chemicals, although the evidence is slim. The most solid evidence of the effect of TSCA on obtaining more adequate information about new chemicals, namely the data submitted with the PMNs, indicates that the act has not had any significant benefits in this respect, except to give us, for the first time, a definition of the universe of new chemicals and a way of tracking them if such tracking seems desirable.

Existing Chemicals. The benefits of TSCA with respect to existing chemicals can more easily be divided between obtaining adequate information and controlling unreasonable risk. Three parts of the act relate primarily to obtaining information about existing chemicals: the testing provisions of Section 4, the substantial risk notices under Section 8(e), and the information collection and record-keeping requirements under the other subsections of Section 8.

To date, no information has resulted from the Section 4
authority to require industry to test specified chemicals if
certain criteria are met. No test rules under Section 4 have
become final because of several interrelated reasons. Between
1976 and 1980, OTS construed the Section 4 criteria as being
extremely difficult to meet, requiring among other things a
comprehensive review of the literature on the chemical under
consideration. Inordinate amounts of time and money were spent
trying to compile the necessary information. A court suit
brought by the Natural Resources Defense Council resulted in
OTS re-thinking its assumptions and agreeing to an accelerated
schedule for proposing testing rules. But the emphasis was on
proposing rules, so making the rules final became secondary.
The new administration has abandoned legally binding rules
altogether, putting the emphasis on voluntary testing by
industry. Whether the voluntary approach will be effective
remains to be seen.

Section 8(e), which requires industry to notify the EPA
Administrator if it obtains information indicating that a
substance presents a substantial risk, has been a fertile
source of information about chemical hazards. More than 400
substantial risk notices have been filed. However, the
connection between information and action is unclear with
respect to the notices. I do not know of any action, either
regulatory or voluntary, by either industry or EPA, that has
been taken as a result of an 8(e) notice. Undoubtedly, there
have been some industry actions. The apparent lack of
attention by the government is more troublesome, and the recent
"100-day report" by OTS identifies the need to make response to
8(e) notices a matter of higher priority.

Section 8 of TSCA contains several other important
information-gathering provisions. The only one that has been
implemented is the initial inventory of existing chemicals
required by Section 8(b), but the experience with the inventory
is sobering and instructive. A number of people, myself
included, fought hard to have EPA collect more information from
manufacturers than simply the names of the chemicals in
commerce. In retrospect, we won the battle but lost the war.
Information about production volume and sites was collected,
but the information processing ability of the agency could not
manage the data received. The data are sufficiently confused
so as to be practically useless. Two lessons can be learned
from this experience: first, that the government needs to
become more skillful in collecting and processing data; second,
that there is always a tendency to ask for more data than can
be constructively used.

The current regime in EPA seems to need no reminding about
the second lesson. In an effort partially motivated by the
desire to reduce the amount of incoming information it has
sharply narrowed the purposes for which information will be

collected. One of the basic themes of the 100-day report is the adoption of a sort of triage strategy which gives priority only to information collection efforts required by law or political necessity. Attempts to identify new problems or to collect information to establish overall priorities among potential chemical risks are not likely to be pursued.

For those of us who are impressed by how much we do not know about chemical risks, this narrowed focus is disturbing. It also surrenders what should be one of the main benefits of TSCA, the authority to comprehensively review the universe of commercial chemicals and to establish priorities. However, the narrowed focus is understandable and probably sensible if one accepts the condition of reduced resources available to OTS. Under the rule of Reaganomics, looking for new chemical problems is akin to complaining about the lack of good French restaurants in San Salvador. If survival is in question some desirable goals must be sacrificed.

To summarize, there have been some benefits of increased information about existing chemicals as a result of TSCA, although the benefits have been much less than might be expected. For the first time we have a list of the chemicals commercially manufactured in the United States. The substantial risk notices have provided some increased information and have given us a regular system for alerting the government and the public to possible new hazards. Testing of existing chemicals by industry has increased somewhat and is being further prodded by voluntary agreements under Section 4.

Efforts to use TSCA to actually control unreasonable risks of existing chemicals have been almost non-existent. The statute itself banned the manufacture and use of polychlorinated biphenyls (PCBs) except for totally enclosed uses or uses that the EPA Adminstrator found would not pose an unreasonable risk. As interpreted by EPA this ban affected only about 1% of the uses of PCBs. Marking and labelling rules for PCBs may have had some effect in reducing the amount of PCBs entering the environment, but the effect is insignificant compared to the amount of PCBs already in the environment. For the past several years there has been no domestic manufacture of PCBs and probably no importation of them.

OTS has focused its control efforts on two other chemicals in addition to PCBs. Working in conjunction with the Food and Drug Administration, EPA used TSCA's Section 6 to prohibit the use of chlorofluorocarbons (CFCs) as propellants in nonessential aerosol products. An advanced notice of proposed rulemaking under TSCA outlined approaches for restricting other uses of CFCs, but the attempt to deal with other CFC uses has been abandoned by the Reagan Administration.

Much effort was devoted to considering controls on various uses of asbestos, and in December 1979 an advanced notice of proposed rulemaking solicited views on such controls. No rules

were ever proposed, and the new administration does not seem to
have much interest in dealing with asbestos hazards.   A
voluntary program to identify asbestos hazards in schools
resulted in less than half the schools conducting
inspections.   A rule making the voluntary program mandatory has
recently been promulgated.

Trying to analyze why there has been so little control
action under TSCA would require a whole separate paper.   The
reasons include very conservative legal views about the scope
of TSCA in relation to other laws, a tendency in OTS to prefer
analysis to action, opposition by industry to any controls, and
the legal, bureaucratic, and political obstacles within the
government that make taking any action a Herculean task.   For
the purposes of this paper, it is sufficient to say that there
have been almost no significant benefits of TSCA resulting from
controls on existing chemicals.

Costs of TSCA

There are two types of direct costs of TSCA.   One is the
cost to the Federal Government of administering the act.   The
federal costs are not separately identified in the federal
budget but were probably around 80 or 90 million dollars in
1981.   The other direct costs are those borne by industry.   The
industry costs of complying with TSCA are likely to be much
larger than the government costs, although no very accurate
figures are available.   The 1980 Annual Report of the Council
on Environmental Quality estimated that the total 1979 public
and private costs for complying with TSCA were $300 million.

The indirect costs of TSCA may well be more important than
the direct costs.   In particular there has been a good deal of
concern about the effect that TSCA has had on innovation in the
chemical industry.

It is impossible to clearly separate the effects of TSCA
from a multitude of other factors which contribute to changes
in innovation or the economic condition of the chemical
industry.   Changes in the tax structure or the inflation rate,
for example, have much more impact on innovation and industry
R&D than does TSCA.   But the effects of TSCA cannot be isolated
from these other factors.

Several industry-sponsored studies have tried to estimate
the extent to which innovation in the chemical industry has
declined since passage of TSCA.   The Chemical Manufacturers
Association has stated, "The quantifiable costs of PMN
requirements, including completing the PMN form, may themselves
account for a 54 percent decline in new chemical
introductions.   Since a decline in new chemical introductions
of between 71 and 87 percent may have already occurred as a
result of PMN requirements, the direct costs of the PMN process
are apparently having a very substantial impact and must be

considered at least as important as any other factor." (2, pp. 25-6)

The methodology used by these innovation studies is not fully reliable. In fact we will probably never know how many new chemicals were marketed annually prior to 1976 because no one kept track of the number and there is no way that it can be reconstructed retrospectively. But any reduction in innovation is a cost to society that must be considered.

How much of a cost TSCA has imposed by inhibiting innovation is not known. Not only don't we know how the number of new chemicals after TSCA compares to the number before TSCA, we don't know how to value the chemicals that were not manufactured. The social and economic importance of new chemicals varies quite widely, and it is reasonable to assume that the chemicals that were not manufactured because of TSCA costs were on average less valuable products than the chemicals that were manufactured.

In July 1982, EPA, responding to several petitions from the chemical industry, proposed exempting several broad categories of chemicals from the PMN requirements of TSCA. The exemptions would cover more than half the chemicals which have been subject to the PMN requirement. To the extent that TSCA has had an adverse effect on innovation in the industry, most of this effect would be eliminated if the exemption proposal becomes final. Of course the potential effectiveness of the PMN review would also be significantly reduced.

Reduced innovation may not be the only indirect cost imposed by TSCA. The act may, for example, encourage concentration in the chemical industry because compliance with its provisions will be more difficult for small manufacturers than for large manufacturers. It will be very difficult to detect and quantify the impact of TSCA on such changes, but they should not be ignored.

Are the Benefits Worth the Costs?

It would be very helpful to have a neat quantitative balance sheet to answer the question of whether the benefits of TSCA are worth the costs. But, as I tried to show earlier, such a balance is not possible.

Given the uncertainty on both the benefit and the cost side, arguments about where the balance lies are probably a waste of time. It would be far more fruitful to agree on the basic philosophical premise of TSCA and then go on to explore ways in which the benefits of the act can be increased and the costs lowered.

The philosophical basis of TSCA, as I see it, is that the public, as represented by the government, has a legitimate interest in industry decision-making about chemicals--in deciding, for example, that new chemicals that are an unreasonable risk should not be marketed, or that existing chemicals that may pose an unreasonable risk should be tested.

I have stated this premise rather starkly so as not to conceal its implications. It implies, among other things, that government has a legitimate role to play as a representative of the public, and it implies that dealing with the toxics problem cannot be left entirely to the free market. If we can agree on these premises then we can go on to discuss ways to improve the act so as to improve the ratio of benefits to costs. If we cannot agree on these premises then we are reduced to a basic philosophical difference which will eventually be settled by determining which side has the most political power.

I shall assume that there is agreement on the basic premise of the act and proceed to explore some ways in which it might be improved. I will try to do this by proposing some generalizations about the way TSCA has worked which I hope will further illustrate some of the costs and benefits but which will also point the way toward future approaches to regulating toxic substances.

Implementing TSCA

First, I would say that some aspects of TSCA are inefficient, in other words the costs clearly exceed the benefits. The PMN program provides some examples. Don Clay, in the 100-day report, has stated that "the PMN program as it is now conducted is not completely adjusted to the realities of commercial chemical development." This lack of adjustment produces some inefficient results. It generally does not make sense, for example, to review PMNs for chemicals which will never be commercially marketed, but at least some of the PMNs are in this category. We need to examine the whole implementation of the PMN program to reduce the time spent on low-risk chemicals, to improve the information obtained on potentially high-risk chemicals, and to develop reasonable follow-up procedures for new chemicals. The 100-day report represents a good start at examining these questions.

Second, some aspects of TSCA, at least as currently interpreted, may be unworkable. For example, the law has been used in only a very limited way to regulate existing chemicals. We need to explore the extent to which this limited use is due to problems in the statute itself or to other reasons, and, to the extent the problems are statutory, amendments to the law should be considered.

Third, voluntary compliance is more efficient than regulation, but voluntary compliance will not be achieved unless the possibility of regulatory action is real. Any law is dependent for its effectiveness on voluntary compliance, and during the Carter Administration we had continuous demonstrations of the cumbersomeness of trying to rely on regulation. But total reliance on voluntary compliance will be equally ineffective, because if voluntary compliance were the only answer needed the law would have been unnecessary in the first place. If an action is not in a person's self-interest some kind of additional incentive is needed. Regulation can provide that incentive.

Fourth, one of the benefits of TSCA is to provide a sort of "due-process" for both industry and the public. For the industry this means that there will not be a trial by press release for suspect chemicals. The act provides an orderly and predictable process for deciding whether chemicals pose an unreasonable risk. For the public, the due process involves not being excluded from decisions involving chemical risks. It means, I think, that the public has a right to monitor voluntary agreements with industry and that there must be safeguards to ensure compliance with such agreements. If voluntary agreements are to be an adequate substitute for regulations then they must have the same due-process safeguards as regulations. For example, they must assure public access to the information developed under voluntary testing programs. If such safeguards are not provided, voluntary agreements will simply lend support to those who now view EPA as the Industry Protection Agency.

Fifth, the information aspects of TSCA must be closely related to the control aspects, whether control is under TSCA, under other laws, or voluntary. The costs of collecting information can be quite high. The benefits to be gained from collecting it should be known with some precision before the costs are incurred.

Sixth, many, probably most of the key questions under TSCA are not scientific, at least in the sense that there is no agreement among reputable scientists about the answers. This is not a tactful observation to make to scientists, but I think it is important to understand if major mistakes in implementing the act are to be avoided. The key questions, while they are scientific in the sense of being potentially verifiable by empirical observation, are in practice and reality policy questions. How many animal tests provide satisfactory evidence of carcinogenicity, given the state of current scientific understanding, is not primarily a scientific question. It is a policy question about how much risk society is willing to tolerate, and thus society as a whole, not just scientists, are entitled to a voice in arriving at the answer.

Some of these statements lend themselves to clear conclusions about how costs can be lowered or benefits increased. Others are more complicated and will require ingenuity to develop improved ways of doing things. But I think that they can be used as a springboard for improving the implementation of TSCA, for maximizing the benefits to all parties. Regulatory reform can be made to work for both

industry and the public.   To make it do so is the real
challenge we face.

Literature Cited

1. U.S. Environmental Protection Agency, Office of Pestcides
    and toxic Substances, "Priorities for Office of Toxic
    Substances Operation," Jan. 1982.
2. Chemical Manufacturers Association, "Comments of the CMA on
    the Proposed Economic Impact and Draft Regulatory
    Analyses for EPA's Proposed Regulations for the
    Submission and Review of Premanufcture Notices under
    Section 5 of TSCA," March 16, 1981.

RECEIVED September 18, 1982

Summary

GEORGE W. INGLE

Chemical Manufacturers Association, Washington, DC 20037

Perhaps the major conclusion of this symposium is that there are so far no pronounced and unequivocal impacts of TSCA on society and the chemical industry. For several reasons, including incomplete establishment of required regulations by EPA, and the subsequent time-lags for compliance with these regulations, and establishment of means to monitor their effects, further time will pass before these impacts will be significant. Even for the long term, when all regulations will have been put in place and implemented, the complex interaction of TSCA with one or more of the twenty-odd other federal laws concerned with control of chemical compounds will mute the impacts of TSCA itself.

It follows that the next, or third, symposium on this topic might best be held five years later, in 1987, rather than continue the pace of meeting roughly every two and one-half years, as was the case for this second symposium.

This is not to say that there have been no evident effects. The sixteen presented papers described a wide variety of effects. Some of these were well documented. Still others were considered largely speculative, awaiting the accrual of experience to determine their significance.

Many of these concerns are expressed in the last paper, given by Dr. J.C. Davies, who had much to do with the report of the 1970 Council of Environmental Quality. This report was the initiative which, in turn, led to enactment of TSCA, six years later. From this perspective, his survey of the overall costs and benefits of TSCA is meaningful. Because impacts of TSCA on human health cannot be identified accurately and certainly not quantified well, and one negative effect -- that on innovation in the chemical industry, cited by several speakers -- cannot be monetized accurately, only a qualitative evaluation of TSCA is possible now. The costs of notifying new chemical substances prior to their manufacture, aside from uncertainty as to EPA's requirements, were described an an impediment to innovation. Several suggestions were made to reduce these costs by sharpening focus on the fewer higher-risk new chemicals and their commercialization after notification, and reducing concern for low risk chemicals. Similarly, costs for collecting the information required by TSCA need to be reduced by concentrating

0097-6156/83/0213-0225$06.00/0

on that portion of demonstrated value. Such evaluation can rarely be precise. Regardless, since the question of how much risk will be tolerated by society is more broad policy than science per se, such analyses must continue if TSCA's implementation is to improve and benefits are to be optimized for all concerned interests.

The speakers were selected to represent views of the chemical industry and one of its major trade associations, the regulatory agency (EPA, Office of Pesticides and Toxic Substances), an international law firm and consulting contractors serving industry and/or EPA, a major university and one centrally positioned environmental group. Thus, differences of views were expected and were evident. As such, these views should appeal to the broad spectrum of American Chemical Society members.

Eight chapters presented observations and conclusions from representatives of major chemical manufacturers, in addition to a ninth from their trade association, the Chemical Manufacturers Association. The first of these, by Nalco Chemical Company's E.H. Hurst, surveyed EPA's accomplishments and problems, and the industry's responses over the five years since the effective date of TSCA, January 1, 1977. TSCA's aim to fill the gaps among the existing laws controlling selected categories or uses of chemical substances, and to identify and eliminate unreasonable risks due to chemicals, was stated. Even more important was TSCA's "balancing" theme, to reflect economic factors, and social benefits of chemicals in reducing their risks. At the same time, Congress did not define "unreasonable risk" or seven other gradations of risk mentioned in TSCA, but left this difficult task to EPA.

In addition to summarizing very briefly the scope of each major section of the Act, E.H. Hurst contrasted the self-implementing sections that became effective immediately and those sections with later timetables for EPA's rule-making and implementation. In the first class are Sec. 8(e), notices of "Substantial Risks," and Sec. 5, "Premanufacturing Notices" of intent to manufacture or process a new chemical substances. Over 400 Sec. 8(e) notices and some 1200 "PMN's" under Sec. 5(e) have been sent to EPA -- substantial evidence of compliance.

In the second category, that of rule-making by EPA, the point was made that, under the Carter Administration, many of these proposals were judged by industry to exceed congressional intent, and to be unworkable in many cases, or workable but costly and ineffective. One example of this thinking is the resulting trend reported in research and development -- to restrict those activities to existing substances to avoid the extra costs and burdens of premanufacturing notification of new substances. This general position was analyzed in substantial detail by S. Davis, Esq., who described "Chemical Industry Initiatives to Modify TSCA Regulations" to achieve compliance with the statute. This account was drawn from the Chemical Manufacturers Association's publication, "The First Four Years of The Toxic Substances Control Act."

E.H. Hurst's overview introduced several themes pursued by other chemical industry speakers. The Dow Chemical Company's E.H. Blair analyzed the problem of setting priorities for testing the 55,000 existing chemicals listed in the TSCA inventory for their effects on health and the environment. Resources for such testing are not unlimited. A systematic classification was made of these substances by production volume. The 9.5% of these substances which account for 99.9% of reported production were divided further into categories such as organic, inorganic, and polymeric.

The commodities among these substances generally have reasonably complete data for health and environmental effects, although data for certain effects may be missing. For smaller-volume organic chemicals, little or none of these data is available. For those chemical compounds in use for decades, major health and ecological effects would have been published had they been observed in manufacture or use. It was shown that the considerable testing resources throughout the world are dedicated by national or private groups to testing those substances which would be expected to pose risks to health or to the environment. Still, in the United States, with existing screening and case-by-case selection systems, such as that used by the Interagency Testing Committee established under TSCA's Sec. 4, it was forecast that five to ten years of such effort will be needed to fill adequately identified gaps in required information.

TSCA has given further impetus to an orderly testing of existing chemical substances, largely by their manufacturers and generally on a shared-cost basis. EPA's H.M. Fribush showed how such evaluations, and other reporting requirements of TSCA, for metal-working fluids confirmed their contamination with tumorigenic nitrosamines. New water-based formulations, avoiding nitrate rust inhibitors and, instead, using new multifunctional additives, consequently have been developed.

This type of constructive impact of TSCA was viewed somewhat differently by Exxon Chemical Company's C.W. Umland. His review of nearly three years of premanufacturing notification cited evidence which suggests substantial disruption of new chemical development and introduction to the market. This disruption was traced to higher research and development costs at an economically vulnerable point in the life cycle of innovative products. A more appropriate balance between opportunity for economic viability and protection from unreasonable

risk for innovative chemicals is needed. Such improved innovation would be expedited by exemption from notification of well
defined, low-risk categories of new substances. Essentially, the
same argument was made by Muskegon Chemical Company's J.R. Yost
during this symposium. The smaller company frequently makes a
major contribution to the flow of new compounds, but is affected
disproportionately by regulation in general, and TSCA's Sec. 5
on premanufacturing notification in particular. A sharper focus
on the higher-risk chemicals was seconded.

EPA's D.G. Bannerman reviewed these impacts on the market
introduction of new chemicals. He summarized EPA's experience
and analyzed the classes and types of new chemicals, company
size, market areas, and, among other data, the number of
notified chemicals actually reported to be commercialized. He
stressed a new joint industry-EPA program to assist the smaller
chemical companies to comply with TSCA, especially with
premanufacturing notification. This will minimize negative
impacts on product innovation without reducing the effectiveness
of EPA's assessment of risks of new chemicals.

These specific, and other broader, concerns in corporate
compliance with TSCA's requirements were discussed by Diamond
Shamrock Corporation's D. Harlow. He described how corporate
structures and procedures, including those for research and
development, for companies of all sizes, have been impacted by
TSCA. These impacts are generally positive in that they reflect
greatly increased awareness, resources and responsiveness to
questions of chemicals' effects on health and environment. These
benefits are seen to be in balance with their costs, expressed
in the increased costs of products and services.

One particular area of this response is the management of
information required by TSCA. Monsanto Company's C. Elmer and
J.R. Condray itemized these requirements, and reviewed the
status and implications of each. Some unanticipated benefits
derived from the mandated burden were emphasized. EPA's requirements for information are not yet complete: further growth is
expected. While EPA's Chemical Substances Information Network
(CSIN) is recognized as valid in concept, there is concern that
its scope may be expanding that originally envisioned for information submitted to EPA under TSCA. In addition, attention must
be given to the identification and maintenance of the reliability level of the information reported, stored and extracted for
use under TSCA.

One continuing problem area in management of this and similar data required by other national systems is their confidentiality. Procter and Gamble's J. O'Reilly presented a perspective
on this issue, which is dealt with explicitly by TSCA's Sec. 14
and has so far been managed well by EPA. Problems are arising in

the differing treatment of confidentiality by other nations
concerned with similar controls of chemicals within the juris-
dictions of the European Economic Community (EEC) and the
Organisation for Economic Cooperation and Development(OECD).
Aside from troublesome differentials in concepts and procedure,
there is the concern that denial of protection of trade secrets
elsewhere will weaken their security under USA laws and
regulations. A proposal was made by Mr. O'Reilly to adopt some
of TSCA's procedures in other countries' rules, to obtain
greater uniformity in protecting justified confidentiality
worldwide.

The protection of trade secrets is only one of the differences among the increasing number of national laws controlling
chemical risks. B.Biles, Esq., described these variations among
the U.S. and corresponding European laws. These laws are in
various stages of development in compliance with the Sixth
Amendment to EEC's 1967 Directive on classifying, packaging and
labeling dangerous substances. The term "harmonization" is used
often to describe the goal of efforts to reduce these
differences and the resulting burden of stifled innovation,
non-tariff barriers to international trade, and inefficiencies
in allocating scarce scientific, technical resources of government and of the chemical industry. Biles opined that it is
unlikely these differences will ever be reconciled by a common
approach to regulating new chemicals. Nor will major multilateral agreements be reached on fundamental regulatory issues.
Rather, accumulating experience -- which is relatively short in
Europe -- including responses to domestic and intra-EEC economic
and political factors, will "fine tune" these laws through
serial regulations, policy statements and administrative decisions. These changes may produce greater consistency. The desire
for harmonization itself will not be the primary stimulus.
Meanwhile, manufacturers should plan initially, especially in
notifying new chemicals, to comply with two or more laws, and
anticipate only a slow movement toward consistency. In a section
on prospects for the future, Biles ventures several forecasts
for harmonizing notification and other requirements over the
next decade. This time frame suggests another reason for
postponing until 1987, or later, the next ACS symposium on this
subject!

Thus far, the twelve views presented have been by
representatives of the chemical industry itself, or of the
industry by EPA, or by an outside counsel. This diversity is
fitting not only because of the industry's primary responsibilities under TSCA, but also because of the longer period of time
for these responses to show effects outside the industry.

As stated earlier, the unique nature of TSCA is its intent
to balance the benefits of a chemical substance with its adverse

effects on health and environment across the spectrum of public health and private interests. This purpose contrasts the goals of earlier environmental legislation to eliminate risks to health and environment, without regard to maintaining such benefits. Accomplishing this balance of effects has forced the continuing development of concepts and practices for distinguishing between reasonable and unreasonable risks. How will EPA reach decisions on this scale, and how will the industry and the public understand and support EPA's decision process?

To answer these and related questions, Decision Focus's D.W. North described a quantitative decision analysis for choosing among alternatives whose consequences are uncertain. This analysis rests on judgmental probability to quantify uncertainty. After reviewing the concepts of quantitative risk analysis and of cost-benefit analysis to show how decision analysis relates to them, an illustrative case study was presented drawn from a specific EPA project on perchloroethylene. At no point was it alleged that such analysis provides easily the "right" regulatory decision under TSCA. Complexities and uncertainties abound. What these disciplines do provide is a decision framework for illuminating these complexities and uncertainties for analysis by all concerned interests. In this way, insights are given, with improved communication and probability for consensus, more so than controversy.

The ultimate goal of such balancing of factors and decision making under TSCA is reduction of risks to health -- public and occupational -- and to the environment. As mentioned earlier, because of limited action by EPA under TSCA, related to the environment (aside from restricting manufacture, use and disposal of polychlorinated biphenyls, banning non-essential uses of chlorofluorocarbons, and from issuing a final rule on disposal of wastes of 2,3,7,8-tetrachlorodibenzo-p-dioxin), only impacts on health were discussed, by SRI's M.J. Lipsett. He stressed the practical limitations on measurement of TSCA's impacts to health. The inability to isolate TSCA's effects from those of other laws and the relative insensitivity of epidemologic studies to long-term effects of low level exposures are major impediments. Add to this list the preventive nature of TSCA. It follows that the best indicators must be estimates, extrapolated from animal tests, and then with much uncertainty.

Regardless, it is concluded that TSCA may have had some indirect health effects. Manufacturers of chemicals have increased their awareness of chronic hazards in the workplace, so that occupational exposures are likely to be lower than in the past. This greater awareness is due not to TSCA alone, but

to other influences such as product liability litigation, OSHA
regulations, and persistent and greater media coverage.

Overall, it must be recognized that preventing unreasonable
risks is harder to implement as a policy, and to measure as
progress, than achieving percentage air reductions in air emissions of particuular pollutants. This explains, in part, why
implementation of TSCA has concentrated so far on gathering
information rather than on controlling chemicals.

At best, these comments suggest that, perhaps in the
future, there will be monitoring systems sufficient in number,
specificity and reliability to determine if each of the many
activities generated by TSCA is cost-effective. If this study is
to be accomplished, there will have to be new leadership to
manage toxic substances by selecting and integrating the several
disciplines required. UCLA's R.L. Perrine proposed that an
important route to this is educational. He described a four-year
doctoral program to create "environmental chemical professionals" trained in the several disciplines, and in problem
solution, not merely informative research. Application of
knowledge across environmental, technical and societal fields is
stressed in the UCLA program, which started in 1970 and now
produces about sixteen "Environmental Doctors" each year. This
is clearly a long-term approach. Expectations for the future are
high. A critical policy question in this educational effort is
this: "Where do we place the balance point in a choice between
education at the doctoral level designed to meet the needs of
society, and education that most conveniently adapts to the
needs for understanding and research publication of university
faculty?"

This long-term view of the costs and benefits of TSCA, why
these cannot be quantified and how they may be managed better,
characterizes the chapter by the Conservation Foundation's J.C.
Davies. His critique of cost-benefit analysis contrasts that of
D.W. North. Regardless, he lists separately some benefits and
costs of TSCA. As to new chemicals, there is very limited
evidence that unreasonable risks may have been averted from a
few new chemicals. In general, aside from systems for defining
new chemicals and monitoring these, if needed, he finds no significant benefits in this area of TSCA.

As to existing chemicals, while much information has been
collected, what is most important is realizing the need to be
more selective and skillful in collecting and processing data.
Section 8(e) has been a fertile source of information on substantial risks, but only recently has EPA recognized the need to
respond to these risks. EPA's failure to use its authority to
require industry to test selected chemicals is similarly

disparaged; so is EPA's current emphasis on voluntary testing by industry. In fairness, however, it is recognized that the record will show if this change in concept will be effective. Overall, the narrowing of focus to concentrate on chemicals of greater risk is perceived as essential to survival under the current restrictions in EPA's resources. No significant benefits are seen resulting from TSCA's controls on existing chemicals.

Direct and indirect costs are compared; public and private costs are estimated at 3.5-4 times those for EPA in 1981. Among the former is loss of innovation. While several studies of this factor have been made for the industry, their reliability is questioned, due in part to lack of sound data prior to 1976. No mention was made of economic trends affecting corporate expenditures for research and development, or of trends in the maturation of industrial chemistry itself. Other indirect costs, such as concentration of manufacture within the industry, may result from costs of compliance, especially for smaller manufacturers. These factors were not compared with extrinsic factors, such as shifts in feedstock supply and commodity manufacture from the United States to other countries.

Davies finds it impossible to determine if the benefits are worth the costs. He suggests that if we can agree that TSCA provides for public participation in industry decision-making about chemicals, then we may be able to define ways of decreasing TSCA's cost-benefit ratio. EPA's "100 Day Report" points the way to reduce time on low-risk new substances and to improve analysis and monitoring of high-risk chemicals. But Davies thinks that current interpretation of regulating existing chemicals is unworkable, and dependence on voluntary compliance is excessive. Unless the "due process" provided by TSCA will apply equally to voluntary actions, then the cynics who now view EPA as the "Industry Protection Agency" may be right. Also, TSCA's costly acquisition of information must be meshed more closely with controls, under TSCA, other laws or voluntary, to obtain commensurate benefits.

The final suggestion, and possibly least palatable to the ACS membership, is to recognize that most of the critical questions involve more policy than science. Better science is always needed but it is a policy question about how much risk -- chemical or otherwise -- society will tolerate. Thus, society as a whole, not only scientists, are entitled to participate in developing the answer.

The symposium ended on this philosophic note. Perhaps its sequel -- in five or ten years -- will be able to present more definitive evidence -- from government, chemical industry and other sources -- as to the merit of TSCA.

RECEIVED August 10, 1982

INDEX

INDEX

Indexing and production by Florence H. Edwards and Anne G. Bigler
Jacket design by Martha Sewall

Elements typeset by Service Composition Co., Baltimore, MD
Printed and bound by Maple Press Co., York, PA